Can Science Fix Climate Change?

Can Science Fix Climate Change?
A Case Against Climate Engineering

MIKE HULME

polity

First published in 2014 by Polity Press

Polity Press
65 Bridge Street
Cambridge CB2 1UR, UK

Polity Press
350 Main Street
Malden, MA 02148, USA

ISBN-13: 978-0-7456-8205-1
ISBN-13: 978-0-7456-8206-8(pb)

A catalogue record for this book is available from the British Library.

Typeset in 11 on 15 pt Adobe Garamond
by Toppan Best-set Premedia Limited
Printed and bound in Great Britain by Clays Ltd, St Ives PLC

For further information on Polity, visit our website: www.politybooks.com

CONTENTS

ACKNOWLEDGEMENTS

I would like to thank my colleagues Martin Mahony, Rob Bellamy and Helen Pallet for reading and commenting on early drafts of this book, and also two anonymous readers of the full manuscript, who made valuable suggestions for clarification and improvement. For what remains I take full responsibility. At Polity I would like to thank Emma Longstaff for commissioning this title, Manuela-Maria Tecusan for her meticulous copy-editing and the production and marketing team, especially Neil de Cort, Elen Griffiths and Ginny Graham. The index was compiled by Bill Johncocks, who, as always, performed the task with great professionalism and diligence.

CDR   carbon dioxide removal
ENMOD   the Environmental Modification
             Convention
EPSRC   the Engineering and Physical
             Sciences Research Council
ETC   the Erosion, Technology and
             Concentration action group
IPCC   Intergovernmental Panel on Climate
             Change
SPICE   Stratospheric Particle Injection for
             Climate Engineering project
SRM   solar radiation management/sunlight
             reflection methods
UNESCO   the United Nations Educational,
             Scientific and Cultural
             Organisation
UNFCCC   United Nations Framework
             Convention on Climate Change

Between 1997 and 2009 the belief persisted among most world leaders that climate change was a problem that could at least be contained, if not completely solved. In December 1997 the Kyoto Protocol was signed at the 3rd Conference of the Parties to the United Nations' Climate Convention, committing industrialised nations to reduce their greenhouse gas emissions by 5 per cent over the next 15 years. The senior British political negotiator John Prescott famously declared: 'This is a truly historic deal, which will help curb the problems of climate change.' Yet 12 years later, in December 2009, the hope of reaching a 'fair, ambitious and binding' follow-on deal at the 15th Conference of the Parties in Copenhagen ended in failure, rancour and disillusionment. The multilateral negotiating process now moves to the 21st Conference of the Parties in Paris in December 2015, where the nations of the world have agreed, yet again, to sign a treaty that should commit all nations to start reducing emissions by 2020. Few are holding their breath.

Yet the risks associated with anthropogenic climate change continue to be encountered somewhere between our direct experience of the world and our science-fuelled imaginations. As global emissions of greenhouse gases – especially carbon dioxide – have soared in the 17 years since the Kyoto Protocol was negotiated, ocean heat accumulates, Arctic sea ice shrinks, heatwaves intensify, the ocean acidifies and sea level rises. Curiously, however, global air temperature – the favoured headline indicator of climate change – has barely increased. The warming trend in global surface air temperature between 1970 and 1998 was 1.7°C per century; between 1998 and 2012, just 0.4°C per century. At the very least, this suggests that knowledge of the complete dynamics of the planetary system is not yet complete.

Yet there are few concerns in the world today that can match the global salience and cultural reach of climate change.[1] After more than a quarter century of scientific investigation, public debate, political negotiation and policy development, climate change and its contested causes remain for many – although not all – a hot-button issue. All of human life is now lived out not just in the presence of a physically changing climate, but in the new discursive and cultural spaces that have been created by the idea of climate change. It is as though all human practices and disputes now can be – now have to be? – expressed through the language and symbolism of climate change.

So photography, cartoons, poetry, music, literature, theatre, dance, religious practice, architecture, educational curricula, and so on now use climate change as a medium of expression. Books dealing with climate change now appear at the rate of more than one per day in the English language alone, up from less than one per week a generation ago. And political disputes about landscape aesthetics, child-rearing, trade tariffs, theology, patents, extreme weather, justice, taxation, even democracy itself, find themselves inescapably caught up in the language and argumentative spaces of climate change. Climate change has become a new condition around which human life takes shape; we 'feel there might not be *any* narrative whose meaning we cannot re-evaluate in relation to climate change'.[2]

So here is the paradox of climate change. While worldwide awareness of climate change as a matter of concern has grown during these last two decades, there remain few problems that are as politically intractable. Top-down multilateralism orchestrated through the United Nations – the so-called Plan A – has abjectly failed to reign in global emissions of greenhouse gases. And politically and culturally engineered behavioural changes within nation states have brought about only marginal adjustments to the underlying drivers of energy consumption and land use change. The world's expanding energy supply continues to be drawn mostly from fossil fuel sources, and the growth in material consumption was only halted – temporarily – by the global recession of 2009–10.

If neither global diplomacy nor changes in human behaviour can solve climate change, a new argument is now being advanced: the world needs a Plan B for regulating climate and, more importantly, science and technology can deliver one. By intervening directly in the heat flows from the sun to the Earth's lower atmosphere, it is deemed possible by some that a thermostat for the planet could be created. The plausibility of this global thermostat relies upon innovative atmospheric modification technologies, guided by reliable scientific knowledge of how the planet's climate works. The most frequently advanced intervention would involve injecting millions of tonnes of sulphur gas into the high atmosphere – so-called stratospheric aerosol injection, just one in a family of sunlight reflection methods. Climate control would then become possible not just for our cars, our buildings and our homes but – it is claimed – for the planet as well.

But *can* science fix climate change this way? And, even if it can, is it a solution we should pursue? In this book I outline the reasons why I believe this particular climate fix – creating a thermostat for the planet through aerosols injected into the stratosphere – is undesirable, ungovernable and unreliable. It is *undesirable* because regulating global temperature is not the same thing as controlling local weather and climate. It is *ungovernable* because there is no plausible and legitimate process for deciding who sets the world's temperature. And it is *unreliable* because of the law of unintended consequences: deliberate intervention in the atmosphere on a global scale will lead to unpredictable, dangerous and contentious outcomes. I make my position clear: I do not wish to live in this brave new climate-controlled world. In Aldous Huxley's novel *Brave New World*,[3] his ironic utopia was brought about by totalitarian engineering of the human subject – 'Yes, everybody's happy now.' I seek to show that, if we promote technologies for a designer climate, an equivalent pathological utopia brought about by totalitarian engineering of the planet would 'likely' be the result.

In Chapter 1 I introduce the idea of geoengineering – more specifically, the science and technology of injecting aerosol particles into the stratosphere to cool the planet. The emergence over recent years of the idea of engineering the world's climate is described, along with some reasons for the increasing attention the idea has been receiving. The notion of a 'climate emergency', which is frequently used as a justification for this type of technological pre-emption, is criticised.

The following three chapters then develop different arguments against the idea that science can fix climate change by developing a thermostat in the sky. Chapter 2 argues that the design of a global thermostat is undesirable: extreme weather and climate change pose risks to humans and the things they care about, but seeking to minimise these risks by regulating global temperature is misguided. Chapter 3 argues that such a thermostat would be ungovernable: world agreement on the desirable temperature setting is unattainable, and the mere attempt to reach such agreement is likely to unsettle international relations. An imaginary scenario for the year 2032 illustrates the point (see Box 3.3). And Chapter 4 argues that the thermostat would be unreliable: even *if* such a technology could be created and governed, the unintended consequences would multiply the humanitarian, political, legal and security troubles facing the world.

In conclusion, Chapter 5 offers a different view of the kinds of problems that climate and its changes present to mankind. This reframing suggests a different role for science and technology: not to try and fix climate change through planetary-scale engineering, but to serve human-inspired goals of dignity, freedom and creativity. It is not climate change that is the ultimate threat to human well-being. It is the lack of virtue. This view requires greater humility with regard to the limits of scientific knowledge than is implied by planetary engineers – and also a greater exercise of compassion and justice in the deployment of that knowledge. It recognises that human beings will never be masters of the planet in the way in which some of the new climate designers seem to be suggesting.

ONE

Imagining an Engineered Climate

Techno-fixing the climate

The Nobel Prize-winning English scientist Michael Beard is on the point of unveiling to the world a won-der-fix for climate change. His solar-powered technology for splitting water into oxygen and hydrogen gas is about to demonstrate its vast potential for cheap, unlimited hydrogen-based energy:

> The day after tomorrow a new chapter would begin in the history of industrial civilisation, and the Earth's future would be assured. The sun would shine on an empty patch of land in the boot heel of south-west New Mexico . . . and the storage tanks would fill with gas.[4]

But Beard's excessive and self-indulgent past behaviour catches up on him. His sins and misdemeanours lead to a tragicomic denouement as his cuckolded adversary takes a sledgehammer to the salvation technology,

1

leaving nothing but wrecked machinery. Beard's dream of a techno-fix for climate change is crushed amid spiralling debts and the chaos of torn human relationships.

The above scenario is from the novel *Solar*, Ian McEwan's comic allegory of climate change. Finding a cheap and sustainable way of creating clean electricity without relying on fossil fuels is one way in which science and technology are frequently imagined to be able to fix climate change. And not just in fiction. At various times hopes have been raised that anthropogenic climate change can be solved through a transition to a hydrogen-based economy, through the wonder-technology of nuclear fusion or through a German-style *Energiewende*: an energy revolution based on large-scale deployment of solar, wind and other renewable energies.

But in recent years another discourse has emerged, in which science and technology are offered as a 'fix' for climate change, this time through the development and deployment of so-called geoengineering technologies (see Box 1.1). These rather eclectic technologies are united in their ambition to deliberately manipulate the atmosphere's mediating role in the planetary heat budget. They aim to do one of two things: either to accelerate the removal of carbon dioxide from the global atmosphere; or else to reflect more sunlight away from the Earth's surface and so to compensate for the heating of the planet caused by rising concentrations of greenhouse gases.

The possible realisation of geoengineering technologies – especially the idea of injecting sulphur gas into the stratosphere – was given a great scientific and psychological boost in 2006. The Nobel Prize-winning Dutch scientist Paul Crutzen wrote an influential article, titled 'Albedo Enhancement by Stratospheric Sulfur Injections: A Contribution to Resolve a Policy Dilemma'.[5] In it, he suggested that the time had come for Earth system scientists to seriously research this method as a backstop technology for limiting climate change – the ultimate techno-fix for climate change. Crutzen was no Beard. He was real rather than fictional; and he was rather more disciplined in his personal life than was Michael Beard. And he had won *his* Nobel Prize for work of great societal benefit. Rather than the abstract 'Beard–Einstein Conflation' with which McEwan's eponymous celebrity scientist had endowed the world, Crutzen's recognition in 1995 was for work conducted with two colleagues on the formation and decomposition of stratospheric ozone. This work had informed the 1985 Vienna Convention for the Protection of the Ozone Layer, which in turn led to the later global agreement to phase out production of the most damaging chloroflourocarbons that had been destroying stratospheric ozone at high latitudes.

Crutzen was careful to state in his 2006 article that injecting aerosols into the stratosphere was 'by far not the best solution' for climate change. Yet his intervention in 2006 spawned a series of research workshops, conferences, networks, studies, assessments and governmental hearings into the possibilities and risks of such a technology. The science reporter Eli Kintisch attended one such meeting in 2007, organised by the University of Calgary and Harvard University. 'Should scientists study novel ways to alter Earth's climate to counteract global warming?' asked Kintisch.[6] The 50 elite researchers at the meeting concluded 'yes', but only after agreeing that 'the road to understanding the science is fraught with booby traps and that deliberately tinkering with the climate could make the problem worse'.[7] One of those scientists in attendance, David Battisti from the University of Washington, was seriously concerned about the meeting's outcome. Kintisch reported: 'After speaking on the phone with his wife from his hotel room, Battisti confessed: "I told her this meeting is terrifying me".'[8]

In this chapter I explain how this idea of creating a techno-fix for climate change by artificially reflecting sunlight has gained such plausibility in recent years and why some scientists such as Battisti might be privately terrified of the idea. But first I introduce the range of planetary intervention technologies that are often grouped under the label 'geoengineering'.

What is geoengineering?

In controversial public debates it matters how terms are defined and understood. The term 'geoengineering' is a rather eclectic catch-all expression. The first use of *geoengineering* in the context of climate change was made by the Italian physicist Cesare Marchetti in 1977, in a rather obscure article titled 'On Geoengineering and the CO_2 Problem'. Marchetti was proposing to capture carbon dioxide emitted from power stations and to bury it either in the deep ocean (his preferred solution) or else underground, in the geological strata. Thirty years later, in the 2006 article referred to above, Paul Crutzen used the labels 'geoengineering' and 'climate engineering' interchangeably, to describe his proposal for stratospheric aerosol injection. More recently, the conventional definition of geoengineering has come to be this: 'the deliberate, large-scale manipulation of the planetary environment in order to counteract anthropogenic climate change'. This definition is taken from the Royal Society's 2009 report *Geoengineering the Climate: Science, Governance and Uncertainty* (Royal Society 2009), which divides geoengineering technologies into two types: *solar radiation management* (SRM) and *carbon dioxide removal* (CDR).

The former set of technologies seeks to offset global warming by reducing incoming solar radiation; and it proposes to achieve this reduction by reflecting more sunlight back into space. Hence these reflecting technologies are sometimes called *sunlight reflection methods* (abbreviated with the same acronym SRM). These technologies include placing mirrors in near-Earth space orbit; injecting tiny sunlight-reflecting particles into the stratosphere (Crutzen's proposal); whitening low-level marine clouds by spraying seawater into them; and whitewashing dark urban infrastructures – roads, car parks, roof tops. The latter set of technologies seeks to remove carbon dioxide from the atmosphere and to secure it in long-term reservoirs. These sequestration technologies include ocean iron fertilisation, soil biochar, and carbon capture and storage. The latter is a process whereby carbon dioxide is captured either from the free atmosphere or from the waste flues of fossil-fuel powered stations. Box 1.1 offers a brief description of the main geoengineering technologies.

Box 1.1 Some Geoengineering Technologies

The first four technologies described below are sunlight reflection methods and leave atmospheric concentrations of carbon dioxide unaffected. In contrast, the four technologies in the second group remove carbon dioxide from the atmosphere and sequester or isolate the carbon in a variety of reservoirs, in effect reversing the process of fossil-fuel combustion.

Stratospheric aerosol injection This sunlight reflection technology seeks to mimic the cooling effect of huge volcanic eruptions. Millions of tonnes of hydrogen sulphide or sulphur dioxide would be artificially injected into the stratosphere, where these gases would oxidise into tiny sulphate aerosol particles, just a few tenths of a micron in diameter. Surviving here for a few years, these aerosols would scatter sunlight back into space, thereby reducing lower atmospheric heating. Other types of particle for injection have been suggested, but the favoured technology would use sulphur gas.

Marine cloud brightening Rather than ask sulphate aerosols to do the job of reflecting sunlight, this method would seek to accomplish it by whitening low-level clouds. The idea would be for mobile ships to spray jets of seawater droplets into the marine lower atmosphere. These would provide condensation nuclei on which cloud water droplets would condense, brightening the marine clouds that occur naturally over large parts of the world's oceans. Brighter clouds would reflect more sunlight back into space, cooling the lower atmosphere.

Orbital mirrors This method of sunlight reflection would insert trillions of tiny metallic reflectors into near-Earth orbit, to reflect the incoming solar radiation back into space before it ever entered the Earth's atmosphere. No chemicals would be involved, but the costs are thought to be prohibitive.

Urban whitewashing Another sunlight reflection method operates over built environments rather than over oceans. White surfaces reflect more sunlight than dark surfaces, and so the idea is to increase the reflectivity of urban environments by whitewashing roofs, roads and pavements. It would be most effective in cities in sunny regions in the summer; but, given the limited surface area suitable for this purpose, the overall planetary effect of this method would be rather negligible.

Biochar (charcoal) When organic matter – wood, leaves, straw, manure – decomposes in a low- or zero-oxygen environment, it produces charcoal, a highly concentrated carbon solid that is resistant to further decomposition. When performed at high temperatures, the decomposition process is known as pyrolysis and effectively traps the carbon that the plant matter originally absorbed from the atmosphere. The resulting biochar can be added to soils to improve their properties and increase agricultural productivity. Under the right conditions of production and management, biochar can lock up carbon dioxide for hundreds of years.

Ocean fertilisation There are large parts of the ocean where algal growth is limited by nutrient deficiencies. By artificially adding large quantities of nitrates, phosphates or iron, this technology would seek to boost algal production and in this way increase the absorption of carbon dioxide from the atmosphere. When these algae die, they sink to the ocean bottom, thus locking the carbon in the deep ocean. This process is slow, likely to be expensive, and has unknown consequences for marine life.

Carbon capture and storage These technologies offer the most obvious way of directly reversing the process of carbon dioxide release through fossil-fuel combustion. They chemically scrub carbon dioxide out of the air and then store the carbon underground, in deep reservoirs. The technology is most plausible – and efficient – for use in the flues of fixed power stations; this is the so-called 'post-combustion capture'. But it is also possible to operate systems for capturing carbon from the ambient air anywhere in the world, most likely in locations close to sources of cheap energy and with nearby suitable underground reservoirs.

Enhanced weathering Carbon dioxide in the atmosphere reacts with carbonate and silicate rocks to produce bicarbonate rocks. These enter the soil and result in the long-term sequestering of carbon dioxide. This process happens naturally, but only very slowly. Speeding it up, for example by crushing carbonate rocks and spreading them out over large surface areas, would be a form of geoengineering. Although the process itself is of low risk, the scale of mining, transport, energy and land area involved in such an operation if enough carbon dioxide is to be removed from the atmosphere would be prohibitive.

There are other ways of subdividing geoengineering technologies, based on different sets of attributes. Thus a distinction can be made between encapsulated and un-encapsulated technologies, depending on whether the method is modular and contained (e.g. air capture and space reflectors), or whether it involves material released into the wider environment (e.g. aerosol injection or ocean iron fertilisation). The scale of the intervention forms the basis for another typology: roof-top whitewashing and soil biochar are both locally implemented, whereas free-air carbon capture and aerosol injection are trans-territorial in their effects. Robert Olson has suggested another category of climate engineering technology with his neologism 'soft geoengineering'.[9] These are technologies that are local in scale, analogous to natural processes, reversible and cost-effective and have no (or few) side-effects and multiple benefits. It is not at all clear, however, which proposed geoengineering technologies – if any – meet all of these criteria.

In this book I challenge the use of sunlight reflection methods to fix climate change, especially the idea of stratospheric aerosol injection, which has risen to prominence in recent years. This particular geoengineering technology – this particular scientific 'fix' for climate change – requires close and critical scrutiny for three reasons. First, it has been claimed to be the easiest and cheapest of geoengineering technologies to implement (see Box 1.2). Second, there is a growing number of scientists who are wishing to start field (atmospheric) experiments to test various aspects of the technology. For example, the Canadian geoscientist David Keith outlines one approach to such experimentation in his recent book *A Case for Climate Engineering* (Keith 2013). And, third, of all the technologies listed in Box 1.1, the environmental, social, ethical and political risks associated with aerosol injection are the most troubling.

The arguments I offer do not readily transfer to the carbon dioxide removal technologies listed in Box 1.1, nor are they necessarily relevant for this group of technologies. They are not my target.

Feeding the imagination

Humans have long desired to take control of their weather and climate and to improve it, just as they have with all other aspects of their physical environment. The big difference between improving the soil, water, plants and animals upon which humans rely and improving the weather, however, is that, for most cultures and for most of human history, the weather has been understood as the 'domain of the gods'. Weather and its behaviour have been taken to be controlled by non-human cosmic powers. These powers – gods, spirits or forces – may be appeased, offered sacrifice, invoked through prayer or, more recently, studied through various forms of rational enquiry. But they are not powers that most humans – even in their moments of greatest hubris – have supposed could be overthrown or superseded through human ingenuity.

This has not stopped some from trying. As historian Jim Fleming recounts in his book *Fixing the Sky*, in the nineteenth century America offered a culture conducive to turning the entrepreneurial mind to ideas of controlling the weather. Robert St George Dyrenforth was one such entrepreneur. He believed that rain could be artificially induced by triggering explosions in mid-air, and in the last decades of the nineteenth century he secured federal funding to test out his balloon-based explosive technologies. Dyrenforth's efforts were unsuccessful and, as Fleming explains, 'he sits in the sincere but deluded category of those who became overly enthusiastic about a single technique or theory'.[10] And in 1965, 40 years before Paul Crutzen's intervention, the United States' Presidential Science Advisory Committee advocated research into the 'possibilities of deliberately bringing about countervailing climatic changes' to offset others, which might be caused at some point in the future by rising concentrations of carbon dioxide.[11] These possibilities extended to changing the reflectivity of land surfaces in deserts and on ice caps, to diverting the courses of mighty rivers – as was undertaken at the time by Soviet Russia – and even to damming the Bering Straits between Russia and the USA so as to alter the climate of the Arctic.

But over the last seven or eight years interest in climate control technologies has been rekindled by

Crutzen's 2006 article. Writing three years before the failed international climate negotiations at Copenhagen in December 2009, Crutzen pointed to the largely unsuccessful attempts of nations to reduce their emissions of greenhouse gases. He argued that this political failure justified, even necessitated, the serious consideration of more drastic – that is, 'extreme and forceful' – geoengineering technologies. His argument was further fuelled, from 2005 onwards, by the widespread adoption of the new metaphor of 'climate tipping points', first by climate scientists and then by policy and environmental activists. This language of tipping points gave new vigour to the notion of putative climate emergencies (an idea that I examine in a later section) and thereby bolstered calls for research into sunlight reflection methods.

After Crutzen's intervention, a small number of influential scientists and technologists began to lead and steer the scientific and public debates on climate engineering and started to influence new science policy and funding initiatives. This group of mostly North American and UK male scientists was labelled the 'geoclique' by Eli Kintisch in his book *Hack the Planet* (Kintisch 2010). These individuals – for example David Keith and Ken Caldeira in North America; John Shepherd and Paul Crutzen in Europe – gained considerable prominence in media reporting of these new technologies, and also great influence in scientific research networks, at parliamentary hearings and on governmental committees. New funding streams for research began to flow and the idea of geoengineering gained visibility in both conventional and new social media.

And so, in Anglophone countries at least, the idea of using sunlight reflection methods to offset some or all anthropogenic global warming moved into the mainstream. *The Independent* newspaper in the UK ran a bold front-page headline on Friday 2 January 2009: 'Climate Scientists: It's Time for Plan B'. With Plan A failing – the Kyoto Protocol and international climate negotiations to control greenhouse gas emissions – the Plan B metaphor quickly gained traction. Plan B was 'the notion that to save the planet from climate change, we must artificially tweak its thermostat by firing dust into the atmosphere to deflect the sun's rays . . . or perhaps even by launching clouds of mirrors in space'.[12]

Later that spring, President Obama's new science advisor John Holdren, in an interview with Associated Press, stated that he believed that geoengineering options such as stratospheric solar reflection methods needed looking into. 'Such actions could not be taken lightly', he said, even though 'we might get desperate enough to want to use it [= geoengineering]'.[13] (Holdren was later forced to point out that these were his personal views and not those of the president's new Administration.) Scientific institutions and self-organising research initiatives also stepped in, to give weight to this new technological possibility. The Royal Society published its first comprehensive assessment of geoengineering technologies on 1 September 2009, although its recommendations were divergently reported in the British media. The following day's newspaper headlines included 'Investment in Geo-Engineering Needed Immediately, Says Royal Society' from *The Times* and 'Royal Society Warns Climate Engineering "Could Cause Disaster"' from *The Guardian*.

The seriousness with which the scientific community was now taking the possibility of climate engineering was further illustrated a few months later. During the week of 22–6 March 2010, about 175 experts from 15 countries convened at the Asilomar Conference Center near Monterey in California. Their purpose was to develop guidelines – on the basis of the nascent Oxford Principles (see Box 3.1) – to control the emerging field of geoengineering research, especially stratospheric aerosol injection. By selecting this venue, these Earth system scientists self-consciously drew parallels with the 1975 meeting at Asilomar, at which molecular biologists laid down a voluntary regulatory framework for research into recombinant DNA.

Thus, in the space of a few short years, the possibility of intervening directly in the atmosphere to modify the planet's climate had moved beyond being the object of abstract speculation and obscure theoretical research. It was now the focus of serious scientific and public policy, and even of legal deliberation in North American and in some European countries. The American Meteorological Society adopted a policy statement on geoengineering the climate system in the summer of 2009, and numerous reports flowed out of NGOs, policy think-tanks and academic research institutes. During the winter of 2009/10, Congressional hearings on geoengineering governance were held in Washington DC and also in London, by the House of Commons Select Committee on Science and Technology. And in 2011 the Bi-Partisan Policy Center in Washington released the report of its 'blue ribbon task force' into climate remediation technologies.

As with any new or imagined technology, linguistic innovation was necessary to communicate across different social worlds the novel risks and opportunities involved in sunlight reflection. Metaphors play a crucial role in such public framing, and three master metaphors have been especially prominent in the media that report on climate engineering: 'the planet is a body', 'the planet is a machine' and 'the planet is a patient/addict'. As a 'body', the planet needs a sunshade or a sun cream; and sunlight reflection methods can provide these. As a 'machine', the planet can be fixed or controlled by designing a thermostat (I will look at this metaphor more closely in Chapter 2). And, as an 'addict', the planet needs radical treatment: we may be addicted to burning fossil fuels, but a replacement substance is a quicker fix than spending a lot of time and effort weaning the addict off the drug. The late American climatologist Stephen Schneider used the analogy of methadone: 'If you have a heroin addict, the correct treatment is hospitalisation, therapy and a long rehab. But if they absolutely refuse, methadone is better than heroin'.[14]

The master argument frequently used in much of this media reporting is that geoengineering is the only option if we want to avoid a planetary catastrophe, 'a climate emergency' – and that sunlight reflection methods are the most plausible technologies to achieve this radical goal (see Box 1.2). As reported by the convenors of a UNESCO meeting that considered geoengineering in November 2010, these methods are 'proposed as an emergency stop-gap to prevent the climate from passing critical tipping points of change'.[15] If we are facing a climate emergency, then radical measures are called for. We need science to 'fix' climate change. But what *is* a climate emergency, and why might some people be keen to develop this rhetoric of emergency?

Box 1.2 Four Reasons Why Researching Stratospheric Aerosol Injection Is Advocated

IT IS SIMPLE AND FAST-ACTING By mimicking large volcanic eruptions, which inject huge quantities of aerosol into the stratosphere, it can be argued that stratospheric aerosol injection is a relatively simple – and natural – technology. It is also fast-acting: within months from the injection, reductions in incoming solar radiation would be achieved and reductions in surface temperature would become detectable.

IT IS CHEAP Although reliable economic costs cannot be established before the precise technology is developed,

---

speculative estimates suggest that stratospheric aerosol injection is one of the cheapest forms of geoengineering, if not the cheapest. The Royal Society report of 2009 suggested this; and one estimate indicates that US$250 billion invested in aerosol injection might be enough to offset the global warming for the coming century. This would be a fraction of the cost of conventional climate mitigation, and such estimates have been seized on by advocates.

IT OFFERS A BRIDGE TOWARDS A LOW-CARBON ECONOMY This is a view espoused by Clive Hamilton in his book *Earthmasters: The Dawn of the Age of Climate Engineering*.[16] Although not in favour of stratospheric aerosol injection, Hamilton suggests that the only justification for it would be to make it easier to transition to a zero-carbon energy economy. It is a technology that might 'buy us time'.

WE NEED IT TO KEEP ALL OPTIONS OPEN This is a commonly expressed view among Earth system scientists and environmental technologists, and it was articulated by Obama's science advisor John Holdren in 2009. This position argues that climate change is such a concerning problem that no response should be ruled out; the technology should be researched so as to become available, should it ever be needed.

Climate emergencies

A decade ago the sociologist Craig Calhoun wrote persuasively about the new cultural conditions of the twenty-first century, which have given rise to a world of emergencies. 'A discourse of emergencies is now central to international affairs'.[17] Emergencies are dramatic, crisis-fuelled constructions, and they can take on many shapes and guises: humanitarian, public health, security, environmental, political. Emergencies and the language and imagery they conjure shape the way we see the world, our place in it and the possibilities and limits of human agency. More importantly, emergencies demand a response, often quick, often radical; 'the international emergency, it is implied, both can and should be managed', says Calhoun.[18] Emergencies cannot be ignored, and therefore they provide a justification for action.

It is important to recognise this emergency-shaped world we now inhabit when reflecting on how and why a climate emergency discourse is used as a justification for radical techno-fixes for a changing climate. Placing the human relationship with (global) climate into the 'emergency' category immediately changes the scope of the types of actions that may be justified and the actors who may legitimately undertake them. Most commentators invoke a *future* climate emergency as justification or warrant for research into sunlight reflection methods and their possible deployment. Here are some examples of how the climate emergency frame is used:

- *From a journalist* If climate change 'will be abrupt and catastrophic . . . it is necessary for world authorities to respond quickly and in unison to the threat, and geoengineering may be a last resort'.[19]
- *From the Arctic Methane Emergency Group* 'There are two general, very large feedback processes in the Arctic that definitely will increase as global warming continues. One is melting Arctic ice and the other is emitting Arctic methane. The loss of ice will definitely increase the emission of Arctic methane to the atmosphere, which makes the Arctic sea ice meltdown the big planetary emergency'.[20] And the group uses this argument as a justification for developing sunlight reflection technologies.
- *From an economist* Sunlight reflection methods 'may be the only human action that can cool the planet in an emergency'.[21]

- *From one of the geoclique scientists* 'What happens if in 2040 or 2060 temperature increases are so high that crops are failing throughout tropical regions and billions of people are threatened with famine? We'd better try to understand if there is something we could do, because there's no other way [than geoengineering] to realistically stop the Earth from warming during the course of this century.'[22]

There are many problems with this climate emergency framing and with the practice of invoking it in order to justify radical techno-fixes for climate change. Most obvious is the problem of defining, detecting and announcing a climate emergency. What exactly *is*, or what *could* be, a climate emergency, and who is authorised to define one? At what scale would the emergency have to be? Presumably at a global scale; but then what political institution is authorised to declare a global climate emergency? (In the scenario introduced in Chapter 4, I outline an imaginary account of how this may develop under the auspices of the United Nations.)

It is here, I suggest, that the connection emerges between the tipping point metaphor and climate emergencies. For the Arctic Methane Emergency Group, a climate emergency should be declared once a tipping point has been crossed, as in the case of Arctic sea ice meltdown. But others might suggest that a climate emergency be declared pre-emptively, as soon as the slow-moving planetary system is committed to crossing a putative tipping point some time in the future. Given the inertia of global biogeophysical systems, some argue that human perturbation of the atmosphere and land has already committed the world to crossing future tipping points. In which case we are already living in emergency times, and drastic techno-fixes are called for today.

Such climate emergency scenarios play on people's fears. Portraying the climatic future in melodramatic terms heightens the levels of anxiety and pessimism to which some members of society are prone. Deep-seated human fears about the unknown future are fuelled by the prospect of some cataclysmic breakdown in planetary function, for example along the lines depicted in the 2004 movie *The Day after Tomorrow*. As I have shown, the declaration of an emergency demands action and intervention, but it is hard in such emotionally charged circumstances to conduct rational discussion and to reach well-informed decisions.

The danger is that depicting the future in terms of climate emergencies becomes a self-fulfilling prophecy.

The language of emergency motivates and justifies certain forms of action in the world, not least technological, political, legal and social. As Calhoun argues, invoking an emergency 'is not merely a description of the world, more or less accurate, but an abstraction that plays an active role in constituting reality itself'.[23] If a climate emergency is announced, then urgent and radical action clearly must be needed. The quickest and easiest response will be brought into play without worrying much about the political, ethical or legal consequences. Democracy can be put on hold: you can almost hear the words 'because of the climate emergency . . . we must inject these aerosols into the stratosphere today; the emergency demands it'. Technology will fix the emergency; the strongest will prevails; politics is bypassed. But at what cost? This is a question I explore in Chapter 3.

It is also important to follow Calhoun's broader argument about the world of emergencies in which we now live. There is a close relationship, historically, between designations of emergency and the ever ready presence of the military to 'assist' in the response. This has been seen in many cases of humanitarian emergency where the results of military intervention – however well-intentioned – have been ambiguous at best and destabilising at worst. More obviously, the same troubling associations are found in the context of international security emergencies. (It is noteworthy that many of the proposed delivery systems for aerosol injection co-opt military technology: missiles, artillery, aircraft, barrage balloons). Calling down climate emergencies to justify radical techno-fixes for climate change may be an attractive political strategy for some, but it carries considerable risks. These risks are examined more closely in Chapter 4.

Finally, it is worth asking whose interests are furthered by talking up prospects of a climate emergency. Oxford University scholars Clare Heyward and Steve Rayner have drawn attention to how the language of tipping points and that of climate emergencies have co-evolved to create a new political strategy for securing action on climate change.[24] This discourse is a key element in what they call a new variant of green millenarianism, in which specific courses of action are prescribed by environmentalists to 'save the planet'. But it is not just environmentalists who adopt this strategy. It also appeals to those with a different worldview, those who are more likely to conceive of the planet as a machine amenable to control engineering. Included here may be certain scientists and those working in some private corporations, conservative think-tanks and military agencies.

Metaphors of agency

What the analysis above shows is that very different conceptions of nature may inform and guide different protagonists in their desire to engineer a global climate. And this is where my answering the question 'Can science fix climate change?' takes me beyond science and into philosophy. I showed earlier how most human manipulation of the environment – through the development and deployment of more or less intrusive technologies – has been motivated by a desire to improve our physical circumstances: to increase yields, to domesticate animals, to harvest resources, to ward against 'flood and famine' and other threatening vagaries of the physical world.

But 'improvement' is only one description of what humans do when they impose their will and artefacts on the physical world. There are many other metaphors of agency, which have been used to justify and guide human environmental practice through the ages – protecting, controlling, caring, mastering – metaphors that reflect very different understandings of the natural world, of divine agency and of human responsibility. Take the case of ecological restoration, an idea that has emerged in conservation science since the 1980s. This concept refers to the practice of restoring degraded, damaged or destroyed ecosystems and habitats through deliberate human intervention and action. But what exactly is it that is being restored? Ecosystems are certainly not being restored to their pristine, pre-human condition. Therefore different approaches in this strand of conservation science focus on restoring processes, functions or services. Whether or not this is 'improvement' depends on whether one's philosophy of nature is preservationist, conservationist or developmental.

When it comes to imagining an engineered climate, many different metaphors of agency are used. The most obvious among them is, of course, engineering itself; geoengineering is to engineer the Earth. The etymology of the word 'engineer' (from the Old French *engigneor*, which comes in turn from the late Latin *ingeniare*, 'to devise' – itself a derivative of the classical Latin *ingenium*, 'nature', 'innate capacity', 'natural talent') connects together the ideas of skill, craft, invention and design with the notion of inborn qualities and talents, as in the word 'ingenious' (or Latin *ingeniose*, 'wittily'). So talk about engineering the climate appeals to the idea of an ingenious invention and design, maintained through skilled craftwork.

The metaphor of engineering refers to a process by which the atmosphere is re-designed. But what about the normative purpose of the engineered climate? What is being accomplished through the act of engineering the climate? The metaphors of agency adopted in geo-engineering debates fall broadly into two families. In one family are those metaphors that resonate with pres-ervationist and conservationist philosophies of nature, for example restoring the climate, preserving, naturalis-ing, perfecting, repairing, caring, healing, protecting, stabilising or remediating it. The other family contains metaphors that are more developmental in orientation: managing the climate, civilising, mastering, controlling, disrupting or colonising it. That two contrasting fami-lies of metaphors of agency feature so prominently in discussions about the technologies of sunlight reflection confirms again the divergent ideologies and interests that are in play.

The extent of the human desire to intervene in the global atmosphere to bring about an altered climatic state reveals what is thought to be the appropriate – legitimate or morally sanctioned – roles for humans to play in relation to the physical world. These are questions that have been reflected on and argued over – even fought over – since humans developed a tech-nological culture different from that of any other primate. The difference today is that the question involves alterations to the atmosphere that are literally global in scale and that impinge on the future of all 7.25 billion people on Earth – and the many more yet to come.

Summary

In this chapter I have described the growing realisation that human activities are inadvertently altering the physical working of the climate system on a planetary scale. In recent years the anxieties provoked in this way have driven cohorts of elite scientists, engineers, policy advocates and environmentalists to contemplate a radical techno-fix for climate change. The technologies that have gained most attention and that many have argued are the simplest, cheapest and fastest to deploy are those collectively known as sunlight reflection methods. And, among *these* technologies, injecting aerosols into the stratosphere to 'cool' the planet by mimicking the role of large volcanic eruptions is the one most researched.

The rhetoric that has been developed to provide justification for such technological adventures, most notably the language of climate emergencies, needs careful examination, as do the various metaphors that are being used to imagine and communicate aerosol injection technology to wider audiences. Different metaphors of agency are used to describe this project; some present it as a restoration and protection of the climate, others as mastery and colonisation of the skies. These metaphors tell us much about how we view ourselves and our relationship with the non-human world, our abilities and limitations, our responsibilities, fears and aspirations for the future. The irony is that the brave new world of sunlight reflection methods and designer climates appeals to such different ideological constituencies: on the one hand, those who can see the climatic future only in terms of tipping points and climate emergencies; on the other, those who see this extension of technology into the management of the atmosphere as the inevitable next stage – the only possible next stage, in fact – in the story of human development.

In the next chapter I investigate further these anxieties about future weather and the desire for a stable climate. Some of the scientific preconditions that have made it possible for people with different worldviews to think seriously about creating a thermostat for the planet are explained. I develop my argument as to why such a techno-fix for climate change is undesirable.

TWO

Designing a Global Thermostat

Optimal climates

Humans have long sought ways to make their weather more agreeable. Our distant hominin ancestors, so we are led to believe, walked out of Africa in search of more favourable weather. They successively colonised and abandoned land masses to the north and east as climates waxed and waned through the Quaternary Era. Writers in classical antiquity frequently valorised certain climates, but only after the latter had been purged, in the imagination, of their most disagreeable components. These optimised climates happily coincided with the temperate climates of the eastern Mediterranean, well away from torrid equatorial and frigid polar zones. And medieval Europeans in the fifteenth century were well able to articulate their own idyllic climates, as in this description of the imaginary Land of Cockaigne: 'There is no heat or cold, water or fire, wind or rain, snow or lightning, thunder or hail. Neither are there storms.

Rather, there is eternally fine, clear weather . . . It is always a wonderfully agreeable May'.[25]

By the eighteenth century, European colonisation and new technologies for land improvement were offering the prospect that such climatic dreams might become reality. The French philosopher Georges-Louis Leclerc, count de Buffon, recognising people's growing imprint on the physical forms and processes of the planet, could write that humanity will be able to 'alter the influence of its own climate, thus setting the temperature that suits it best'.[26] (The idea of the thermostat, then, was born at least a century before the world's first electric room thermostat – which was invented by the American Warren S. Johnson in 1883.) Draining swamps and clearing forests for agricultural development were activities seen as part of the project of climatic improvement, to which American settlers and European colonists of the tropics believed they were called.

But there is another tradition of perfecting the climate. Rather than migrating in search of more attractive climates or trying to manipulate the physical climate so as to make it suit human preferences, adaptationists sought to organise their way of life to suit the weather. The historian William B. Meyer tells the story of one such enterprise in New York State: the Oneida community's quest for meteorological utopia in the middle of the nineteenth century. The Oneidans' religious beliefs undermined prevailing ideas that human attributes were passively determined by climate and promoted instead the project of moulding social life so as to accommodate the vicissitudes of the weather. An 1852 editorial in their community newspaper *The Circular* 'dismissed "the idea of imputing evil and mischievous power to the rain and wind and simple elements" as patently inconsistent with Christianity'.[27] The Oneidans' view was that 'bad' weather was as much a consequence of poor planning and social organisation as it was the result of a capricious agent, be that God or nature. As the later aphorism observes rather tritely, there's no such thing as bad weather, just the wrong clothing.

Why are these stories about past efforts to create a perfect climate – through migration, engineering or behavioural adjustments – relevant to my argument? They are relevant because the current aspiration to engineer global climate echoes some of these tales. Most noteworthy perhaps was the ambition of the French socialist visionary Charles Fourier. Fourier was concerned about the deteriorating climate of the 1840s, and among the utopian projects of his American disciples was 'the regulation of the seasons, the moderation of temperatures, and the control of climates, in such a way as to have them always the most favourable'.[28] Fourier's global atmosphere was to be 'serene and genial', rather like the eternally 'agreeable May' of the Land of Cockaigne.

To bring about 'the control of climates' today, the sunlight reflection technologies of the new climate engineers need a control variable to which they can be applied. This variable needs to be one that is ambitious enough to offer attractive prospects for relevant, significant parties; and it needs to have both scientific credibility and political traction. Conveniently, the evolving science and politics of climate change in recent decades have offered just such a variable: global surface air temperature.

In this chapter I explain how global temperature came to be adopted as the dominant metric for revealing climate change. Representing climate change through the language of global temperature is rhetorically powerful – scientifically, politically and culturally. But it is also dangerous, as it offers too easily the imagery of a thermostat and the illusion of planetary control. Climate engineers are proposing and researching technologies that would 'manage' this variable, much as the householder seeks to manage the comfort of her home or car by turning the thermostat. I briefly explain the most favoured of these sunlight reflection methods – stratospheric aerosol injection – and outline the dangers of reducing weather and human welfare to a function of global temperature.

'No more than two degrees'

In the last quarter century we have become used to talking about the world's temperature. But, unlike in the situation observed by Samuel Johnson – that two Englishmen meeting and talking first about the weather 'are in haste to tell each other what each must already know, that it is hot or cold, bright or cloudy, windy or calm'.[29] – to our modern-day conversationalists it is not at all self-evident whether the planet today – this very day – is hot or cold: this is beyond our immediate senses. Taking the world's temperature is a tricky and delicate operation (see Box 2.1).

Nevertheless, the end of each year now sees a flurry of media releases from the handful of research groups that calculate global temperature, giving their early estimate of what the year's average global temperature will likely be. It is much like the eagerly awaited quarterly GDP or inflation statistics, which purport to diagnose the health of the economy. A ranking usually accompanies the temperature estimate – 'warmest yet', 'in the top ten warmest', 'coolest since . . .'. Climate modellers also orient themselves by this quantity. It is even possible to conduct some debates over the veracity and utility of climate models by using solely the vocabulary of global temperature. And the illusive equilibrium climate sensitivity – a measure of how sensitive the planet's climate is to doubling the atmospheric concentration of greenhouse gases – is expressed in terms of change in the global temperature. Estimates of its value range typically from 1.5° to 4.5°C and have changed little in over 35 years of scientific inquiry. Contrarianism, too, finds the language of global temperature useful. Some commentators, still in doubt about the implication of human activities in climate change, appropriate this index in order to claim, for example, that the world shows no warming trend since 1998.

Box 2.1 Taking the World's Temperature

It is no small matter to take the temperature of the planet. A thermometer can be stuck under the human tongue or into an animal's anus; but there is no easy or quick method for taking the planet's temperature. Attempts to gather temperature and other meteorological measurements on

---

regional scales date back to the early nineteenth century, but the first effort to reduce these data to a single index that described 'global' temperature was made in the 1930s by the British engineer Guy Callendar. Though limited by sparse records and hand calculation, Callendar's published trends from 1880 to the 1930s closely match estimates made with today's methods.

The expansion of meteorological networks after the Second World War, combined with the rise of computational data processing in the 1970s, led to more comprehensive efforts of global temperature estimation in the late 1970s and early 1980s. These efforts were led by the UK and the USA, and the first combined land and marine global surface air temperature index was published in 1986. This was shortly before the first global temperature trends derived from satellite measurements became available. But satellite-derived indices refer to the temperature of the mid-atmosphere rather than that of the surface. Because of this, and because thermometer-based estimates of global temperature commence in 1861 while those derived from satellite date only from 1979, it is the former that remain of greater scientific and public value.

The origins of these efforts to take the world's temperature lay in the interests and needs of theoretical climatologists, in particular those concerned with estimating long-term climate change. The empirical credentials of the thermometer-based global temperature index are impeccable; it is a distillation of hundreds of thousands of observations made by tens of thousands of observers at thousands of meteorological stations all over the globe. But this empiricism is confounding. Precisely because of its global origins, the global temperature itself is an empirical impossibility. It exists nowhere and can be experienced by no one.

In addition to this ontological paradox, the global temperature index is dogged by serious practical challenges. As a representation of an average global temperature, it is unavoidably compromised by uneven and frequently non-existent data from large parts of the globe. Problems of calibration and data smoothing are inevitable when dealing with such a large and unruly data set. Even an apparently simple measurement like surface temperature is complicated. Temperature can vary widely at two, five or ten metres above the surface even in the same location (vegetation in the rain forest, for example, causes a significant range in temperature above or below the canopy). Generating meaningful comparisons between thousands of such varied surface measurements is tremendously difficult and not without controversy.

Despite these data collection and comparison challenges, the global temperature index has answered the desires of climate modellers seeking a simple way to validate their simulations of the Earth's climate. It has also gained the status of a quantity around which science, politics and public discussion of climate change can be organised and contested.

It is not surprising that the headline messages from the United Nations' Intergovernmental Panel on Climate Change (IPCC) have long revolved around the future performance of this temperature index. Projections of future warming out to 2100 have ranged from about 3°C above the late twentieth century average in IPCC's first published assessment in 1990 to a likely range of between 0.3°C and 4.8°C in the most recent assessment in 2013. The putative impacts of future climate change are then indexed against this projected temperature change, with different types and magnitudes of impact attached to each degree increment between 1° and 6°C. Mark Lynas utilised the power of this indexed language by giving his polemical 2007 book the title *Six Degrees: Our Future on a Hotter Planet*.

Given its salience in climate science, it is not surprising that the linguistic repertoire of global temperature is mirrored across wider public and policy discourses, where different levels of temperature rise are deemed catastrophic, dangerous, tolerable or at best benign. And so global temperature has become the favoured metric around which the goals of climate policy have been debated. The European Union was the first political authority to articulate a climate policy objective using such language, and in 1996 it stated that its goal was to limit global warming to no more than 2°C above pre-industrial levels. Thirteen years later this goal was adopted by the international negotiators operating under the United Nations Framework Convention on Climate Change (UNFCCC). The Copenhagen Accord agreed in December 2009 stated that the world's governments 'recognize the scientific view that the increase in global temperature should be below 2 degrees Celsius'. This was confirmed most recently at the eighteenth negotiating session in Doha in December 2012: 'Parties will urgently work towards the deep reduction in global greenhouse gas emissions required to hold the increase in global average temperature below 2°C above pre-industrial levels.'

'Two degrees of warming' therefore stands for danger in the twenty-first century, even as many are claiming that the prospects of staying within the designated safety zone are diminishing. Back in 2008 Robert Watson, the UK government's chief scientific advisor on climate change, introduced the slogan 'mitigate for 2, adapt for 4', using global temperature as the organising device around which policy development should be oriented. The ambition was to limit warming to two degrees, but the pragmatic reality was to start adapting for four degrees. Others have now adopted this numerical reasoning, for example Peter Christoff in his recent book *Four Degrees of Global Warming:*

Australia in a Hot World. But, whether two degrees or four degrees, global temperature retains its dominant position as the ultimate descriptor of the state of climate–humans relations.

The temperature of the world therefore pervades the language of climate change. It is hard to conceive of scientific, policy or popular debates about climate change that are not referenced to this quantity of global temperature. In this respect, global temperature is a 'boundary object' that serves the interests of actors in different communities by allowing intelligible communication across their different social worlds. The construction of global temperature out of a mix of empirical and theoretical research cultures enables the index to have the kind of conceptual tractability that makes it a powerful element in both scientific and policy discussions.

And it seems that governments always desire such simple control variables when it comes to discharging their duty to govern. In his book *Seeing like a State: How Certain Schemes to Improve the Human Condition Have Failed*, political scientist James Scott shows how the emergence of modern government co-evolved with the regimenting, mapping and quantifying of the resources and citizens under its jurisdiction.[30] No enumeration, no government. This mentality – or what Michel Foucault famously called 'governmentality' – enabled what Scott calls the high-modernist projects of the twentieth century, so many of which ended in failure and disillusionment. Along this line of reasoning, GDP – an index of aggregated marketed economic activity – becomes a variable for contemporary governments to maximise, just as global temperature – presented as an aggregate of local weather behaviour – becomes a control variable to regulate.

Although global temperature has emerged in recent decades as this powerful boundary object, achieving both scientific credibility and political legitimacy, as an object of political control it brings both deficiencies and dangers. As I show later in this chapter, global temperature is grossly misleading as an index of the state of climate–humans relations. It obscures most of what matters, in terms of weather, to humans and the things to which they are attached: rain to grow crops, wind to power turbines, cyclones from which to shelter, and the like. And it is also a dangerous idea insofar as it offers the metaphorical power of the thermostat imagery and the suggestion of technological control of the planet's climate. But first let me examine how the putative global thermostat would work.

A thermostat for the world

The geoengineering technology that most prompts the metaphor of a global thermostat is that of stratospheric aerosol injection, which intentionally adds sunlight-reflecting aerosols into the high atmosphere. In his seminal 2006 article on this topic, Paul Crutzen compared the artificial development of the technology with naturally occurring processes in the stratosphere that follow large volcanic eruptions. Such explosions are accompanied by injections into the atmosphere of large volumes of pyroclastic debris and sulphur dioxide gas. Although the debris and dust settle or are washed out within days or weeks, the sulphur dioxide converts through chemical processes into sub-micrometre sulphate aerosol particles. These may be just a few tenths of a micron in diameter, and if the explosive plume injects them into the stratosphere – above about 10–12 km altitude – their residence time may extend to months, or even years. Such tiny particles reflect incoming sunlight back into space, preventing the heat from reaching the lower atmosphere and from warming the Earth's surface. This effect is sometimes referred to as global dimming.

After one of the most explosive eruptions in recent times – Mount Pinatubo in the Philippines, which exploded in June 1991 – between 10 and 20 million tonnes of sulphur dioxide were emitted from the volcano. The gas plume entered the stratosphere, where it was oxidised into aerosol particles before circling the planet, drifting into high latitudes and residing for up to two years. The Pinatubo eruption alone caused the average global surface air temperature to cool by about 0.5°C between 1991 and 1992. The even more explosive eruption of Mount Tambora in Indonesia in April 1815 famously led to the European 'year without a summer' in 1816, an event vividly captured in Lord Byron's poem *Darkness*:

> I had a dream, which was not all a dream.
> The bright sun was extinguish'd, and stars
> Did wander darkling in the eternal space,
> Rayless, and pathless, and the icy earth
> Swung blind and blackening in the moonless air.

The proposal to introduce sulphur dioxide deliberately into the stratosphere therefore mimics this naturally – if sporadically – occurring process. The greater the volume injected, the larger the fall in global temperature will be. The design of this planetary thermostat revolves around a number of scientific and technological questions.

The scientific questions concern where exactly in the stratosphere the material should be injected – in the tropics, at higher latitudes, at what altitude – and what exactly the injected material should consist of – sulphur dioxide, hydrogen sulphide or, as was proposed by the late Edward Teller, other artificially manufactured products. Most importantly, what volume of material should be injected and how frequently? The answer to this last question depends on how much global dimming is deemed desirable or necessary to offset any future warming. For the simple case of offsetting the likely global warming that could result from a future doubling of atmospheric greenhouse gas concentration, injecting somewhere between 1.5 and 5 million tonnes of sulphur per year would seem to be necessary. This is roughly equivalent to a Mount Pinatubo eruption occurring once every 2 to 10 years, indefinitely into the future.

The main technological question concerns how to get the required material reliably and safely into the stratosphere. There are three groups of methods most commonly discussed. These are to use high-flying aeroplanes to dispense the aerosols *in situ* at the required locations and altitudes; to use military cannons or surface-to-air missiles to inject aerosol-containing projectiles into the stratosphere; or to use very tall tethered hoses held aloft by balloons to pump the requisite material 10 km into the sky. It was the proposed testing, in 2011, of a very simple prototype of this last option that provoked the public controversy I describe in the following chapter.

The design of such a thermostat raises obvious questions about its safety and reliability, about its unintended side-effects and about the political dimensions of control and governance. These are matters I will look at more carefully in the following chapters. For now I want to ask just this question: Can engineering the world's climate by using global temperature as the control variable ever secure the intended benefits for humans and the things that matter to them?

Global temperature or local weather?

The thermostats we are most familiar with in everyday life operate in our homes, our offices and our transportation vehicles. Whatever their specific design and their setting, the principle is very simple: a control system monitors the temperature of the enclosed space and switches on and off the heating or cooling apparatus to maintain the ambient air temperature as close as possible to a set value. Although some latitude is designed into the control system to regulate conditions within a given margin of the set-point temperature, the key feature of a single-zone thermostat is that the air temperature is uniform within the design space. Multiple-zone thermostats can of course be introduced to allow different temperatures in different spaces – as in separate room thermostats in buildings – but the key point remains: a uniform thermal environment is maintained within the regulated domain.

Using global temperature as the target variable that aerosol injection technology is seeking to regulate is like adopting a single-zone thermostat for the planet. It *may* be possible to engineer planetary heat flows in such a way that globally averaged surface temperature is maintained within a given range (although I will show that even this is questionable). But this most manifestly does *not* ensure that regional climate systems, within which local weather develops, respond in close synchronisation with the global temperature index. The ubiquitous language of global temperature and of the nominal two-degree safety limit seems to have persuaded some that regulating *this* quantity will ensure benefits – or limit damages – for all. But the relationship between people, weather and their security is intensely local. What matters for humans (and what affects their non-human attachments) is not what happens to global temperature, but what happens to winter snow falling on the

Rocky Mountains, to the south Asian monsoon, to typhoons in the Western Pacific, to the reliability of mild moist airflows from the Atlantic Ocean. The risks and benefits of a changing climate and the weather patterns that follow such change – the things that *do* matter to us – are not reducible to an index of global temperature. Seeking to govern climate–people relations this way seems extremely foolhardy.

The folly multiplies when one realises that an artificial manipulation of the stratosphere is not just about consequences for the stratosphere or for global temperature. The consequences extend to all regional climate systems, and hence to the day-to-day sequences of local weather experienced in particular places. The challenge of stratospheric aerosol engineering is not a one-dimensional control problem – to offset the build-up of excessive heat in the Earth system by reducing the overall amount of sunlight that reaches the lower atmosphere. It becomes a multi-dimensional control problem, in which the degrees of freedom in the system rapidly multiply. Not all outcome variables, not all manifestations of local weather can be optimised, let alone controlled.

This feature of the Earth system to which the technology of sulphate aerosol injection is applied has of course not been lost on some analysts. Using a variety of climate simulation models, scientists have explored the regional consequences, for climate and weather, of using stratospheric aerosol injection to stabilise global temperature. One early study from the Department of Engineering and Public Policy at Carnegie Mellon University in Pennsylvania showed that the ostensible goal of 'restoring late-twentieth century climate' was unattainable for all regions. The response of rainfall to the injection of aerosols was that some regions got wetter (for example India) and some drier (for example China). And, although most regions cooled as the thermostat was 'turned down', some regions actually got warmer. As the authors concluded in a rather understated manner,

> it may not be possible to stabilise the climate in all regions simultaneously using [sunlight reflection methods]. Regional diversity in the response to different levels of [aerosol injection] could make consensus about the optimal level of geoengineering difficult, if not impossible, to achieve.[31]

Another study has shown just how sensitive regional rainfall patterns are to exactly *where* in the stratosphere additional aerosols are introduced. If the latter are loaded into the southern hemisphere stratosphere, the drylands of the African Sahel become wetter; if in the northern hemisphere, they become drier. But there is worse news for those who still dream about engineering an improved climate for the world. The *opposite* effect to the Sahel is observed on rainfall amounts in the semi-arid region of northeast Brazil, the Nordeste. So one could 'improve' the climate of the Sahel and 'worsen' the climate of the Nordeste. Or one could choose to favour the Brazilians. But it seems to be a zero-sum game. Thus the prophecy of the mathematician John von Neumann more than half a century ago is fulfilled: 'Intervention in atmospheric and climatic matters will come in a few decades and will unfold on a scale difficult to imagine at present . . . what power over our environment, over all nature, is implied!'[32]

And these are not isolated results. Nearly all the modelling studies that have simulated the regional effects of stratospheric aerosol injection show that the existing mosaic of regional and local climates ends up being reconfigured. Stabilising global temperature to avoid the danger zone of more than two degrees of warming only ends up *destabilising* regional climates around the world. Land grabs will turn into sky grabs and territorial disputes will extend to the stratosphere, as potential powers vie for control of the thermostat. Far from offering a technology that defuses the imagined climate emergency, the aerosol-fuelled thermostat in the sky will create new security emergencies by offering possibilities to manipulate regional climates. The governance nightmares thus unleashed are the subject of the next chapter; but it is important here to warn that some scientists think that all of this can be safely and sensibly resolved through rational use of their models.

The modelling team at Carnegie Mellon University more recently used game theory alongside their climate simulation models to predict how different political actors would react to the different regional consequences, for climate, of setting the thermostat at different levels. Their earlier work had led the scientists to the sanguine belief that the sort of inequality illustrated above was *not* an insurmountable obstacle in designing a policy for implementing sunlight reflection methods. They then demonstrated this by conducting what they called 'the global thermostat game', in which different 'knob settings' on the thermostat would be chosen by different actors. They outline their necessary modelling assumptions, which include the propositions that their model is accurate (!) and that human welfare under conditions of climate change is a function of only temperature and precipitation. Their global thermostat game shows that, rather than a unilateral implementation of stratospheric engineering, 'a more likely possibility is a strategic multilateral implementation through an exclusive "club" [of nations] that increases benefits to the members at the expense of those excluded'.[33]

However one interprets the results of these types of simulation analyses, they all draw attention to the politically charged nature of the deployment of aerosol injection technology in the stratosphere. This engineering project is not simply about designing a thermostat for a single enclosed space or system, where the public benefits of a regulated global temperature are uniformly distributed and recognised. It is about an instrument of technology that – even *if* it worked perfectly on its own terms – would open up a new frontier in the sky for inter-state rivalry, conflict and disagreement. As if the world has not enough reasons already for such quarrelling. Rather than a return to the Cold War, sunlight reflection methods would presage a new era of Sky Wars.

Summary

My argument in this chapter is that there is an inherent flaw in the operationalising of stratospheric aerosol injection technology that makes its pursuit deeply undesirable, if not dangerous. The imagined thermostat works through agents who decide, on the basis of some scientific calculus of predicted dose and response, exactly how much aerosol to inject, where and for how long. These decisions are therefore informed by planetary-scale calculations of global temperature and global radiation balance. The justification for implementing them – the only justification, as far as I can see – is to regulate, maybe to stabilise, global temperature on the grounds of avoiding or defusing a climate emergency.

But the supposed benefits of such an intervention cannot be simply indexed against either of these two global variables: temperature or the radiation balance. The welfare, in relation to weather and climate, of humans and of the things that matter to them cannot be reduced to such a calculus. The public goods and evils associated with climate are distributed unevenly across the world, according to a bewildering array of constantly changing weather systems and events, which are always mediated by social and political structures and institutions.

The idea that global temperature is a suitable object of governance and one through which the well-being of humanity can be secured is a delusion. It too easily suggests the problematic notion of a global 'we', one that collapses valid and competing interests into a unitary global subject: '*we* must limit global warming to no more than two degrees', '*we* must avert a climate emergency'. The risks of a changing climate do not self-evidently lead to the creation of a global 'we' in which the same common objectives are shared by all. Nor are all interests equally well served by engineering a specific planetary temperature with its attendant shifts in regional climates.

The language of global temperature also lends support to the rhetorical idea, used by some, that a stable climate is a global public good. But a stable global climate does not equate to stable local climates; achieving the former does not guarantee achieving the latter. And, if one insists on using the language of public goods, then stable local weather is a much more significant public good than a stable global temperature. As I have shown above, stratospheric aerosol injection, although ostensibly offering a brute force technological means of securing climate stabilisation, in fact destabilises far more than it stabilises. The ethicist Dane Scott summarises the position well: 'the setting for deliberations over sunlight reflection methods will consist of numerous groups with conflicting interests confronting difficult questions of risk, uncertainty and contingent futures'.[34] These indeed are some of the difficult questions that underlie most forms of human antagonism. The use of methods of sunlight reflection is more likely to exacerbate such antagonisms than to resolve them constructively or peacefully.

'Can science fix climate change through global sunlight reflection methods?' is the question I am answering in this book. Certainly not. The technology of stratospheric aerosol injection offers a double fantasy: first, that the risks of a changing climate can be defused effectively by regulating global temperature; and, second, that the world could ever agree on the level at which to set the thermostat. In the next chapter I show why this thermostat is not just undesirable, but also ungovernable.

THREE

Governing the World's Temperature

Spicing it up

At the end of August 2011 *The Guardian* newspaper in London ran a story trailing the impending announcement of 'the world's first major "geo-engineering" field test'. At a decommissioned military airfield in Norfolk, a team of UK scientists intended to undertake a field trial of a scaled-down version of the balloon-and-hose design for injecting aerosols into the stratosphere. Like all successful media stories, it carried an eye-catching headline – 'Giant Pipe and Balloon to Pump Water into the Sky in Climate Experiment' – and it was an exclusive. But the announcement was never made, and within a month the body responsible for funding the project – named, rather mundanely, Stratospheric Particle Injection for Climate Engineering (SPICE) – had suspended the experiment. Eight months later, in May 2012, it cancelled it completely.

What is interesting about this rather peculiar incident is how it revealed so many of the contentious issues surrounding the development and deployment of stratospheric aerosol technology designed to 'fix climate change'. The letters page of *The Guardian* ran opinions both in favour of the field trial – from none other than the president of the Royal Society, Sir Paul Nurse – and against it. An example of the latter was this opinion offered by Thomas Crowley, a resident of East Lothian in Scotland, who raised the question of legal liability for the consequences of climate engineering:

> The proposal for a giant tether designed to combat global warming seems right on the boundary between 'very cutting-edge' and 'crackpot' science. I hope the designers have factored in a legal team to deal with any peculiar weather – anywhere – occurring as a result of this experiment. It seems virtually guaranteed that someone, and perhaps many, will file lawsuits if the slightest aberration in weather occurs. Before spending millions of pounds developing such a system, the investigators, and the funding agencies, should take a very hard look as to whether they can cope with such issues. I doubt that they can.[35]

Within a few weeks of *The Guardian* story, the Canadian-based environmental action group on Erosion, Technology and Concentration (ETC) wrote an open letter and petition to the UK's secretary of state for energy and climate change. It urged the British government to cancel the experiment, claiming that if the field trial went ahead many governments of the global South and many civil society organisations around the world would conclude 'that the UK is preparing to proceed down a very high-risk technological path'. Accusations were made in some quarters that the scientists involved in SPICE had an undeclared patent application for the technology, and the suspicion raised questions about motivation. Was this indeed research, or was it a prelude to commercial development?

After a month of heated public argument and extensive private discussion, the sponsoring organisation – the country's publicly funded Engineering and Physical Sciences Research Council (EPSRC) – announced on 29 September that they had told the experimenters to delay the field trial. This followed advice from 'an advisory panel' and was 'to allow time for more engagement with stakeholders'. The following May, after the controversy had cooled down, the EPSRC finally announced the abandonment of the experiment, citing concerns about the governance of geoengineering and about the existence of patent applications. They defended their actions throughout the episode as reflecting 'a responsible innovation approach' to the project.[36]

In this chapter I explore some of the questions about responsibility and governance that would be raised if climate change were to be 'fixed' through the use of stratospheric aerosol injection technology. How can the technology be trialled safely before deployment? Under what conditions would it be right to start seeding the atmosphere with aerosol particles? How would nations deliberate and reach agreement on how much climate manipulation is needed? Indeed, is such agreement even necessary, or could nations proceed unilaterally? And whose voices count most in these deliberations? The difficulties of reaching multilateral agreement through the United Nations system with regard to conventional climate mitigation suggest that governing novel climate engineering technologies will be harder still. And such deliberate intervention in the climate system would open the way for liability claims against adverse local weather to be issued through international tribunals.

Governing research

In the case of the SPICE project, scientists eventually agreed that, due to the lack of appropriate governance structures for climate engineering, it was inappropriate to carry out a field trial of their balloon-and-hose delivery system. But it had taken a minor public controversy for them to reach this position. And they, too, like the EPSRC, cited their commitment to adopting a 'responsible approach to innovation' – especially to technologies that have the potential to affect people worldwide. The prospect of a thermostat for the planet certainly falls into this category. No one can opt out, save leaving for Mars. The problem of governing stratospheric aerosol injection technology appears to be two-fold: How should the design, funding and execution of *research* into the technology be governed? And under what governance arrangements should the technology be *deployed* on a planetary scale? These two questions are now examined in turn, although I argue that the answers offered to each are unavoidably interrelated.

The huge advances in technology over the last half-century have forced scholars and politicians into careful reflection about the social processes of innovation. This has helped researchers understand early on in the funding and innovation process the opportunities, risks and uncertainties of emerging technologies. Responsible innovation involves opening up for public scrutiny any consideration of the potential ethical and practical consequences of a novel technology even before it has been developed. For example, in the case of the SPICE project a public consultation exercise was undertaken in the summer of 2011 inviting the participation of 32 diverse members of the British public to gauge reactions to the proposed experiment. Although these citizens were not unduly alarmed by the SPICE proto-experiment itself, they *were* concerned about the lack of regulation of such experiments in general, and also about the absence of any international governance mechanism for ongoing research or possible future deployment.

Scientific research into aerosol injection technology and its effectiveness is unprecedented in its characteristics and implications. Unlike, say, the products of medical and pharmaceutical research, which was the comparative case cited by Sir Paul Nurse in his public intervention mentioned above, aerosols and their injection into the stratosphere cannot be tested on animals. Nor can double-blind randomised control trials of the technology be performed on multiple subjects. There is only one subject: the planet we live on. If field trials – even limited ones, like SPICE – are to proceed, the technology can only be shown to deliver its desired effects if the entire planet is subject to the experiment. This prompts the use of computer simulation models and their attendant uncertainties (an issue I examine in Chapter 4).

This proposed climate-engineering technology is therefore unlike nearly all other novel technologies, where the scales of testing and deployment are either territorially limited or bounded in some other way. The technologies emerging from the genetic modification of organisms, from neurochemistry and from nano-engineering, for example, can in theory all be researched under tightly controlled laboratory or field conditions. Only then are they deployed operationally, and only at limited sites or on restricted subjects. The effectiveness of these technologies is not dependent upon planetary-scale implementation. But the technology of stratospheric aerosol injection is different. Perhaps the closest equivalent is the research, development and deployment of the atomic bomb in the 1940s; indeed some commentators such as Jay Michaelson at the Hebrew University in Jerusalem refer to it as a 'climate change Manhattan Project'.[37] Whereas existing technologies of weather modification are limited in scope and scale (see Box 3.2), aerosol injection technology has the objective of changing weather and climate everywhere.

Researching any prospective technology that is aimed at manipulating the physical world logically implies that there are conditions under which it is believed that the technology could or should be used. If it were universally recognised that there are *no* such conditions, then there would be no point in researching it. (I draw a distinction here between, say, scientific research into the aerosol chemistry of clouds and research into a technology that aims at deliberately changing that aerosol chemistry in order to yield a desired outcome.) The question about whether or not the technology *would* be deployed is often side-stepped by arguing that, unless the research is undertaken, the dangers of using the technology will never be known; any future decision as to whether or not to implement it should be informed by the best possible knowledge of all the risks. But this still begs the question. It implies that there *are* imaginable circumstances in which the technology might be used. (The comparison with the atomic bomb breaks down at this point, since the Manhattan Project was undertaken under conditions of worldwide conflict and later justified through the logic of deterrence; this is not the logic of sunlight reflection.)

How, then, should research into stratospheric aerosol injection technologies be governed? The voices in favour of undertaking such research – whether in the form of computer simulations or by way of field trials of delivery systems or particle injection – are loud and persistent, and they have gained the ear of those governmental authorities and funding agencies that have so far considered the matter. But, rather than recommending unregulated research into the technology, these advocates have suggested a number of governance principles or rules to which all researching scientists and technological entrepreneurs should adhere (see Box 3.1). Needless to say, they resist a moratorium on such research.

Box 3.1 Principles for Governing Geoengineering Research

The Oxford Principles were drafted early in 2010 by a UK-based team of scholars led by the University of Oxford.[38] They emerged as a consequence of the Royal Society's 2009 report into geoengineering and they were built on the deliberations of the Asilomar Conference. They have subsequently been endorsed by the UK's House of Commons Science and Technology Select Committee and by the British government. The principles are:

- Since people cannot opt out, geoengineering must be regulated as a global public good.
- The public must participate in decision-making with regard to geoengineering research.
- Full disclosure of geoengineering research projects and the open publication of all results must be guaranteed.

- Independent assessments of the impacts of any research proposal must be completed.
- Governing arrangements must be agreed upon prior to deployment.

These principles have been designed to apply to the full range of geoengineering technologies, but they are particularly relevant for my discussion of stratospheric aerosol injection. It is questionable whether the SPICE field trial met these conditions. The authors of the Oxford Principles have argued that the principles should be adopted internationally and that broad public engagement should ensure everywhere that geoengineering research has a social licence to operate.

A complementary set of guiding principles have been proposed by North American scientists Ted Parson and David Keith, specifically in relation to stratospheric aerosol injection research. They propose that two thresholds of project scale and risk be defined in order to isolate three categories of experiment.[39] Field trials of a scale above the upper threshold should be banned; scientists should not pursue them and governments would not fund them. Field trials of a scale below the lower threshold would be allowed to proceed with government funding and with some level of public scrutiny. The authors do not specify what should happen to proposals in the middle category, between the two thresholds.

There are other approaches to conducting research into stratospheric aerosol injection. One idea, promoted by Eli Kintisch and David Keith, is to appoint two teams of researchers – a blue team and a red team – both funded publicly. The blue team should be funded to research and develop appropriate technological systems for a global thermostat. It might consist of scientists and engineers who are good at finding technical solutions to problems. The sole objective of the red team, on the other hand, would be to identify as many possible weaknesses in the novel systems proposed by the blue team. The red team might consist of scientists of a more sceptical mind, who are good at identifying scientific or design flaws.

The emergence of these various principles, rules of conduct and proposals reveals the level of concern that many scientists entertain about research into aerosol injection. This unease is due to many of the reasons revealed when the SPICE field trial was aborted in 2011: unknown or uncontrollable side-effects; a sense of unease about a planetary-scale intervention that uses human technology; vested human interests. Jane Long, the director-at-large of the Center for Global Strategic Research at the Lawrence Livermore National Laboratory near San Francisco, summarises these vested interests as comprising 'fortune, fear, fame and fanaticism.'

She warns against their corrupting effects on the research process, as does the Australian commentator Clive Hamilton:

> So how do we prevent the formation of a powerful constituency of scientists, investors and politicians after a quick fix, a lobby that could manipulate the political system to downplay or override serious concerns about safety in order to see its technology deployed?[40]

No matter how cautiously research into stratospheric aerosol injection technology might proceed, no matter by what means field trials, small or large, may be governed, and no matter how effective the red team/blue team dynamics may be, there remains one overriding concern about the promotion of such research: it draws too clean a distinction between researching and deploying the technology. Once one is started on a course of technological development, it becomes increasingly difficult to stop its eventual deployment. The slope becomes increasingly slippery. Those who advocate research into stratospheric aerosol injection and who have developed governance principles for such research also have a responsibility to conceive and demonstrate how the full deployment of the technology would be governed securely and justly.

Governing deployment

The slippery slope argument, applied in this context, is that, once stratospheric aerosol injection technology has reached a certain scale of development, the next stage of development is easily justified. The momentum grows until full-scale deployment eventually becomes unstoppable. The slope is 'slippery' because of technological and sociological 'lock-in' – the outcome of the powerful economic, cultural and discursive dynamics of innovation. The ideology of technological progress is deeply ingrained in contemporary society. There are the vested interests of fortune, fear, fame and fanaticism pushing the technology onwards. And, discursively, the allure of a technological 'solution' to climate change will gain followers as more conventional proposals to reduce the scale of human influence on the climate system are stymied by political inertia and intransigence. In Chapter 5 I will suggest an alternative response to this intransigence.

Full-scale deployment of stratospheric aerosol injection technology is, by definition, planetary-scale deployment; hence at the bottom of the slope lies an operational global thermostat. The philosopher Friedrich Rapp places the slippery slope argument in the context of broader technological development:

the lesson of history is that once the first step has been taken development cannot be stopped; global scientific, technological and economic systems of exchange and competition guarantee that whatever has become standard will very soon spread all over the world . . . for this reason it is easier and safer to stop before getting started than it is to slow development already in progress.[41]

The governance of research cannot and should not be separated from the governance of deployment. If the deployment of the technology cannot conceivably be adequately governed, then the technology itself should not be researched. In which case the question becomes: Can we conceive of governance regimes under which the technology could safely, legally and justly be operated? We might look at other, comparable cases for inspiration. Box 3.2 recounts two different ways in which social and political institutions intervened to offer oversight and eventual regulation of cloud-seeding technologies in the twentieth century. The courts intervened in the case of a civil action; and the United Nations intervened in the case of the militarisation of weather modification. But the characteristics and contexts of stratospheric aerosol injection, in contrast to those of weather modification, suggest that the world would be confronted with much greater degrees of political, legal and technical complexity in designing a suitable governance regime.

Box 3.2 The Governance of Cloud Seeding

An interesting case with which to compare the governance of stratospheric aerosol injection technologies is that of cloud seeding. The first successful development of cloud-seeding technology occurred in the United States in the years immediately following the Second World War. The General Electric Company first claimed to have 'made snow fall' by seeding clouds with dry ice – solid particles of carbon dioxide that created nuclei around which water could freeze and precipitate. Over the next few decades it became commonplace to seed certain type of favourable clouds, with the express purpose of stimulating them to precipitate their water over designated areas. The technology most frequently in use today disperses silver iodide from aircraft that fly through conducive clouds to accelerate the nucleation process.

The USA, Israel, Australia, Iran, China and Indonesia are just some of the countries that have applied this technology on a large scale. For example, the department of agriculture in Texas supports such weather modification over an area covering one fifth of the state, and up to one tenth of the American Midwest has been subject to

---

commercial cloud seeding. China famously claimed to have used cloud-seeding technology to avert rain at the 2008 Olympics, seeding clouds elsewhere to ensure rain did not fall over Beijing. And in 2013 the Indonesian government came under pressure from Singapore and Malaysia to expand its cloud-seeding programme in the direction of creating rain, in order to suppress widespread forest fires whose smoke palls were suffocating these neighbouring countries. Earlier that year Indonesia had used the same technology to try to *reduce* flooding in Java by causing tropical clouds to deposit rainfall over the adjacent ocean.

Although there remains scanty evidence that cloud-seeding techniques reliably alter the distribution or yield of rain, what is of relevance here is how the technology is governed. I will mention two contrasting cases. The first is a court case. An early action was brought against New York City in 1951 by plaintiffs who claimed that flood damage of their property had been caused by a private rainmaker, Dr Wallace Howell. Although the citizens from the Catskill Mountains community were not awarded damages, they won a permanent injunction that terminated further cloud-seeding activities by the City.[42]

A different approach to governing this weather modification technology emerged at the height of the Cold War. Between 1966 and 1972 the US military used cloud seeding in Vietnam. The aim of Operations Popeye and Motorpool was to increase rainfall in North Vietnam and to impede the military operations of the opposing Viet Cong through flooded roads and mudslides. After Popeye and Motorpool were exposed at the end of the Nixon era, the Soviet Union successfully steered a regulatory treaty through the United Nations. Known as the Environmental Modification Convention (ENMOD), this treaty explicitly prohibited the 'military or other hostile use' of weather modification technologies that would have 'widespread, long-lasting or severe effects . . . on any other State Party'. The treaty opened for signature in Geneva on 18 May 1977 and today 85 states have signed, including the United States. It is at the margins of possibility that any deliberate intervention in the stratosphere that knowingly produced a serious change in climate affecting another state might be in violation of ENMOD.

The assumption of many is that the deployment of stratospheric aerosol technology could only be undertaken under the auspices of a multilateral United Nations process. The most successful instance of international environmental diplomacy leading to a regulatory regime is the Montreal Protocol for the Control of Ozone Depleting Substances signed in 1987. On the other hand, one of the least successful multilateral ventures has been the attempt to regulate global emissions of greenhouse gases under the United Nations Framework Convention on Climate Change (UNFCCC). The questions that stratospheric aerosol injection would raise in front of a United Nations multilateral process are as politically complex as those afflicting emissions control, if not more so. They include the following:

- when to implement the technology;
- how much warming to offset and for how long;
- how to deal with international liabilities for untoward regional side-effects;
- the creation of a technical agency for implementation and of an inclusive high-level governing body.

Given the deficiencies of the UNFCCC, it is difficult to envisage geopolitical circumstances in which conditions would be propitious for reaching agreement on these matters. The proponents of the idea of Earth system governance or of a world environmental organisation do not have empirical evidence on their side.

If governance of stratospheric aerosol injection technology is unattainable through the United Nations, then two options remain: a consortium-based regime; or unilateral implementation. The former runs along the same lines as the idealised outcome of the global thermostat game conducted by the scientists at Carnegie Mellon University that I looked at in the last chapter. These scientists concluded that a suitable governance regime might be induced simply by agreeing that anyone who wants to participate in a coalition of nations designed to operate the thermostat can do so. They suggested that this would ensure that 'the global thermostat is not controlled by a few at the expense of others'. Political scientists are much more sceptical. Informal consortiums of nations such as this one would be too unstable to provide effective governance of a technology that, by definition, would extend over decades.

This leaves open only the third option – unilateral action by a single nation. Action of this kind might be undertaken by a 'rogue state' – such as North Korea – operating under a megalomaniac delusion of grandeur. Or perhaps it would be performed by a powerful nation that sought to further its own interests and framed the necessity for deployment in terms of national security; in this case stratospheric engineering would be implemented on account of a perceived *national* climate emergency rather than on account of an agreed global one. But unilateral implementation of stratospheric aerosol injection offers no satisfactory means for its governance. On the contrary, it simply recognises that certain political entities may act on their own, being motivated by self-interest – or possibly self-interest combined with a measure of altruism, if the state involved genuinely believed that the global benefits exceeded the risks.

So we are left with a set of undesirable governance scenarios, one of which is illustrated in the imaginary world around the years 2032 and 2033 (see Box 3.3).

Box 3.3 A Scenario

It is February 2032. Two glacial lake dam bursts in the Himalayas in October killed 35,000 people in Nepal; one million Madrid residents remain relocated in Spanish coastal cities following last summer's intolerable heat and the ongoing water shortages; the Egyptian agriculture sector has collapsed following a tenth year of low-Nile flows, and Lake Nasser stands virtually empty; western Sydney remains derelict and uninhabited following December's uncontrollable wildfires. Six thousand Australians died in the blaze.

Germany – one of the new permanent members of the United Nations Security Council – has put forward a formal resolution for the UN to start the systematic injection of sulphate aerosols into the stratosphere. In her speech to the Security Council, the German foreign minister claims that such a direct climate engineering measure is called for, given the condition of the world's climate and its future prospects. January's inundation of large parts of Hamburg – after an unprecedented North Sea storm and sea surge that caused damage worth €40 billion – was the final catalyst for her motion to declare a global climate emergency.

She goes on to report that last September Arctic sea ice extent shrank to just 17 per cent of its late twentieth-century value; gas monitors in Canadian permafrost have identified rapidly increasing rates of methane release; the moratorium on all new nuclear power plant construction worldwide continues following the explosion at a nuclear reactor in China in 2030. The world's population stands at 8.3 billion and the global temperature averaged over the 2021–30 decade was 15.2°C. This is 0.7°C warmer than in the first decade of the century and 1.5°C warmer than the pre-industrial average. Coal, oil and gas continue to be burned worldwide in vast quantities.

The Intergovernmental Panel on Climate Change (IPCC) delivers a special report for the Security Council on the regional climatic risks of such intervention. On the basis of the latest Earth system models, the Panel recommends how much aerosol should be injected to reduce global temperature by 0.5 degrees over the next decade. It also offers probabilistic predictions of the changes in decade-average regional rainfall that would result around the world from such an injection. It cautiously states that there would be a vanishingly small risk of intensified tropical storm activity as a result.

The 15 members of the Security Council consider Germany's claim of a global climate emergency and also scrutinise carefully the predictions made by the IPCC. In particular, they spend much time weighing the probabilities that the Asian monsoon might be weakened as a result of aerosol injection. Security Council members also argue about how long the initial aerosol injection should continue for: 2, 5 or 10 years. Against a background of vociferous (and at times violent) globally coordinated public campaigns – some in favour, but many against such intervention – the Security Council votes 9–4 in favour, with 2 abstentions. None of the five permanent members exercises his or her veto. But, as a compromise, initial deployment will proceed only for one year, after which a full evaluation will be conducted. The injection of aerosols will be managed by the World Environmental Organisation but monitored by one of the new planetary citizen juries established by Google in 2027.

Aerosol injection starts in November 2032; a new generation of miniature superintelligent drones in the stratosphere control and monitor the procedure. Over the following months protestors attempt to sabotage some of the French military planes that are being used for delivery of aerosols into the sky. Direct-action groups affiliated with the campaigning coalition group Hands Off Mother Earth send up their own aircraft, in symbolic efforts to scrub the aerosols from the stratosphere. By early June 2033, the Russian government comes under pressure to call for a suspension of the experiment; and the pressure comes from several of its newly rich shipping magnates. These tycoons were concerned that the longer sea ice season on account of which the summer opening of several Arctic sea routes had been delayed was a result of the aerosol injection. In the face of public and political protests, combined with the devastating loss of life after a super-typhoon in the Philippines in June, the deployment is halted two months earlier than scheduled, in September 2033.

Climate data are evaluated against the results of Earth system models that simulated an ensemble of counterfactual climates for 2033 assuming no aerosols had artificially been added to the stratosphere. Measured global temperature in 2033 fell from the previous decade-average of 15.2° C to just 14.7°; this made it the coolest year on the planet since 2017. Counterfactual modelling suggests that, without the additional aerosols, the temperature would have been 15.1°C, which indicates an aerosol-induced cooling effect of 0.4°C. This is close to the IPCC's estimate that its designated aerosol injection would cool the planet by 0.5°C. But regional climate anomalies have been large and variable. Apart from the 130,000 lives lost in the Philippines during the most intense cyclone season in the South China Sea in 20 years, of greatest concern was the failure of the 2033 Asian monsoon, which cost the Indian economy $70 billion. Model simulations suggest a 73 per cent probability that the monsoon failure is linked to the aerosol injection. This elicits from the Indian government demands for reparations from the UN.

India – one of the rotating members of the Security Council – and China now trigger an emergency debate at the full UN General Assembly in New York in November 2033, calling for a permanent ban on the deployment of aerosol injection technologies. The scientists of the IPCC argue that one year's data prove nothing about the regional impacts of aerosol injection and point to the substantially reduced global temperature. But, against a background of further global protests led by the new popular civic movements in Brazil, Ethiopia and India, the Assembly passes the Indo-Chinese resolution by a large majority.

Turmoil ensues as two Canadian billionaires unilaterally continue aerosol injection into the stratosphere. They believe the stabilising of the Canadian permafrost during the summer of 2033 was a result of the experiment. Continued aerosol injection offers them a means, they claim, of preventing future regional warming, which in recent years has caused much damage to their oil and gas pipelines in the Canadian Arctic. The Inuit Circumpolar Council brings an action against them to the Supreme Court of Canada sitting in Ottawa, on the grounds that aerosol injection violates the 2010 Nuuk Declaration on Inoqatigiinneq (Sharing Life).

The year 2034 begins with heightened geopolitical confusion and tension. With no further chance of reaching agreement through the UN, Germany and France, urged on by the Coalition of Disappearing Island Nations, decide to continue their own aerosol injection programmes for a further five years, to establish the extent to which such manipulations of the stratosphere can reduce the rate of global warming.

My argument remains that, given the likelihood of the slippery slope – one thing leads to another – those who promote research into stratospheric aerosol technology have a dual responsibility. It is not just principles for the global governance of research that are needed, but also a clear and plausible strategy for how the technology would be governed before and after deployment. Such a strategy would need to identify the 'best possible' – or at least the 'best imaginable' – governance regime for stratospheric aerosol injection. And if this best imaginable regime is not good enough to secure the assent of all interested parties (who these parties are I examine below), then the socio-technical imaginary of the thermostat should be dispensed with. Science will not be able to 'fix' climate change this way.

Given the seriousness of the governance dilemma I have pointed to, it is imperative to multiply the voices that are being heard in debates and initiatives around stratospheric aerosol injection. This is not just about the geoclique – those scientists who have been most influential in raising the possibility of such a climate fix. It must extend beyond experts and government officials to engage all the peoples of the world. Since no one on the planet can opt out of an implementation of the thermostat, procedural justice demands that everyone's voice should be heard when it comes to deciding whether one should be installed and at what temperature it should be set. I therefore ask: Whose voices are absent in the current conversations? Does this matter and why?

Absent voices

Chapter 1 showed how the geoclique has been particularly salient in scientific assessments, media reports and public conversations about sunlight reflection methods.

Its members have overwhelmingly come from North America and Europe. It is also noteworthy that most of the government bodies that have engaged with the putative technology are from these same regions. The danger here is that a very specific set of culturally conditioned human values and ethical arguments come to frame and dominate the public conversation. For example, the Royal Society's 2009 report has been extremely influential in setting out the terms of the debate. But, as the American environmental philosopher Stephen Gardiner has pointed out, the values informing this assessment are those that are salient and well rehearsed in western academies and among western government elites. They will not necessarily be shared by all those on the planet.

There remain significant silent spaces on the planet with respect to the expression of public and political preferences about the technology and the associated vision of a global thermostat. Many forms of non-expert civil organisations and associations have barely begun to express their views: the labour movement, religious faiths, trade associations, women's groups, industrial sectors and business communities, indigenous peoples. There is just a handful of exceptions to this observed exclusion. UNESCO organised a meeting in November 2010 to bring experts together from a dozen countries in order to widen the discourse. And, also in 2010, the Solar Radiation Management Governance Initiative was established jointly by the Royal Society in London, by the Environmental Defense Fund in Washington DC and by the Third World Academy of Sciences. Under their auspices expert workshops to consider the governance of sunlight reflection methods have been held in various parts of the world: India, China, Pakistan, Senegal, South Africa and Ethiopia.

It is also true that the assessment of stratospheric aerosol technology by scientific and technical experts has been extended beyond the usual countries under the auspices of the United Nations' IPCC. In their 5th Assessment Report, the IPCC assessed various aspects of a range of geoengineering options, including sunlight reflection methods. However, even here, important questions remain about which experts and civil representatives get to participate. In June 2011 the IPCC convened in Lima, Peru an expert workshop on geoengineering attended by 50 experts nominated by governments. Half of them came from the USA, UK and Germany, and only 15 per cent of them were women. The campaigning coalition Hands Off Mother Earth argued that IPCC scientists do not have the legitimacy or expertise to determine the suitability of governance arrangements. In an open letter to the chair of the IPCC, Dr Rajendra Pachauri, they insisted that

> the critical question of governance is one that needs to be fully debated by the international community, with *all* interested states, civil society organisations, indigenous peoples and farmers' organisations taking part in a clearly democratic, multilateral, transparent and accountable way.[43]

Indigenous peoples offer one example of how certain interests, far away from the centres of scientific and political power, can easily be marginalised. There are cultures for which the very idea of human 'control' of the climate conflicts with their beliefs and values. As has been argued in other contexts, procedural justice for Indigenous peoples involves free, prior and informed consent and an acknowledgement of their sovereignty over the territories upon which they depend. Such interests and voices need to be given recognition in international discussions about a global thermostat.

Another voice absent from the discussion about whether to research or implement a technology that would have profound consequences for the future of the planet is that of the unborn. We are ignorant about the preferences of future generations, and a virtue ethics position would argue that it is unethical to project our preferences into the future. Of course this consideration could be used to argue either for or against implementation. However, there is one ethical argument that turns out to be against implementation. Through our initiation of a global thermostat, future generations would be bequeathed the dilemma of whether to continue with aerosol injection indefinitely (and bear the ongoing risks) or whether to cease the programme (and cause further climate disruption). We can imagine strong reasons why future generations may wish to choose either option, but logically they cannot choose both – hence the dilemma. Some philosophers such as Konrad Ott would claim that leaving such a dilemma for the unborn is morally repugnant when there are other viable options available.

Summary

In this chapter I have argued that a planetary thermostat in the stratosphere would be ungovernable. Deciding when to implement it and agreeing what the setting should be and how it should be governed in the imaginable future would demand an unprecedented – and simply unattainable – degree of trust and cooperation among the nations of the world. The alternative governance scenarios of voluntary coalitions of interested actors or unilaterally implementing political entities are unlikely to be sustainable and will lead to political rancour and possibly to military conflict.

This problem of how stratospheric aerosol technology would be governed undermines not just arguments for its deployment at some point in the future. It also undermines the case for undertaking research today into the technology and its consequences. It is not an appropriate response to claim that 'we' need better to understand the technology in case 'we' should need to deploy it in the future. I have already shown the impossibility of identifying and agreeing on the conditions of a universal climate emergency that might warrant deployment. Such conditions are political and ethical *before* they are scientific and technical. Agreeing on such conditions demands that expression be given to the multitude of perspectives and arguments about why a particular climate condition might or might not be deemed a universal emergency. This is not a task for scientific experts, even less for engineers. Different interests must be given voice and recognition – from both powerful and powerless nations, from Indigenous peoples, from people of faith and of no faith, and they must be ventriloquised for those who are not yet born, to give consideration to their interests.

This necessary politicisation of stratospheric aerosol technology must not be short-circuited by arguments that defer to scientific claims that humanity must in any case arm itself for the future. A global thermostat advocated, designed and run by technocrats is no solution to climate change worth contemplating. It suggests the possibility of a view from nowhere, the possibility of a metaphorical cockpit for Spaceship Earth in which benign and wise experts manipulate the planetary controls for the betterment of humanity. Salvation from the consequences of our actions will not arrive in this manner.

In the following chapter I extend my argument further and show why trying to fix climate change by using science in this way is not only undesirable (Chapter 2) and ungovernable (this chapter); it is also unreliable. Implementing a global thermostat carries with it the inevitability of unintended risks and establishes the conditions for a perpetual and uncontrollable planetary experiment from which there is no escape.

FOUR

Living in an Experimental World

Biosphere-2

In the searing heat of the Arizonan desert, 30 kilometres north of Tucson and in the shadow of Mount Lemmon, is a 16-hectare research site owned by the University of Arizona. On it stands a large sealed glass structure known as the Human Habitat, covering 1.2 hectares. On each side of the Habitat are two smaller domed structures known as the South Lung and the West Lung. This futuristic complex is Biosphere 2, an Earth systems research facility built between 1987 and 1991 by Space Biosphere Ventures. Biosphere 2 was designed to provide an artificial, closed ecological system for researching and learning about the Earth and its living systems. The idea was to explore experimentally the interactions between humans, farming technologies and the rest of nature by mimicking the Earth in miniature. Biosphere 2 allowed the manipulation and study of a biosphere without harming that of the Earth – the original Biosphere 1. The vision for Biosphere 2 extended to pioneering the use of closed biospheres for space colonisation.

Biosphere 2 had two closure experiments, the first of which ran for two years, from 1991 to 1993. Eight Biospherians were sealed inside the glass enclosure and scientists continually monitored the changing biochemistry of the air, water and soil. The health of the crew was also monitored – by a doctor inside the sealed world and by a medical team outside. Several accounts of this experimental life have been published by former Biosphere 2 crew members. Here is a brief extract from the introduction to an interview with one of them, Abigail Alling:

In September of 1991, wearing futuristic jumpsuits made by one of Marilyn Monroe's clothes designers, four men and four women entered Biosphere 2 . . . closed the airlock behind them and, except for one of the women who left Biosphere briefly to have her severed finger sewn back on after a threshing machine accident, emerged two years later. They had orange skin from the high levels of beta-carotene in their diet. They had since become acclimated to oxygen levels that initially found them with symptoms of high-altitude sickness. And they hadn't carried cash in two years.[44]

A second mission followed in March 1994, but a dispute over the financial management of the project caused it to be ended prematurely. All closed-system experimentation in Biosphere 2 ceased after 1994, and the place reverted to a more conventional biological research station. To prevent it from falling into the hands of housing developers, the University of Arizona bought the site in 2007. It is now a science education centre.

Biosphere 2 offers one way in which to conduct experiments on the (miniature) Earth. Another way is to manipulate planetary functioning by using computer simulation tools. But both these experimental methods are deficient when it comes to testing the effects of stratospheric aerosol injection on the Earth's climate. The only experimental method for adequately testing system-wide response is to subject the planet itself to the treatment. This is what those who promote sunlight reflection methods are in effect proposing.

My argument in this chapter is that the project to re-engineer the world's climate by injecting aerosols into the stratosphere is like the designer experiments of Biosphere 2. But, rather than subjecting eight jump-suited Biospherians to a deliberate experiment, aerosol injection would subject Biosphere 1 to the experiment, together with the 7 billion plus people and 8 million plus species who reside here. I show that, as with any experiment, the outcome is unknown and unknowable. The claimed goal of stabilising the climate and defusing a climate emergency is unattainable. This is a view also expressed by the campaigning coalition Hands Off Mother Earth (HOME) in its open letter of June 2011 to the IPCC:

> The potential for accidents, dangerous experiments, inadequate risk assessment, unexpected impacts, unilateralism, private profiteering, disruption of agriculture, inter-state conflict, illegitimate political goals and negative consequences for the global South is high. The likelihood that geoengineering will provide a safe, lasting, democratic and peaceful solution to the climate crisis is non-existent.[45]

I first examine the potential risks of the aerosol experiment and suggest that many of these can barely be imagined, let alone quantified. Deploying the technology leads to a perpetual condition of intentional experimentality for the planet Earth and its inhabitants. As the unforeseen consequences of the experiment emerge, there will be an infinite regress of further interventions, to compensate for adverse outcomes. Aerosol injection is not simply about stabilising or restoring the global

climate. It is an intervention that has profound repercussions for what we think it is to be human. It would
forever alter our sense of moral duty and ethical
responsibility.

The nature of the experiment

In 1957, in one of the earliest scientific analyses of the
effects of rising concentrations of carbon dioxide on
the global climate, the oceanographers Roger Revelle
and Hans Suess famously described human beings as
'carrying out a large-scale geophysical experiment of a
kind that could not have happened in the past nor be
reproduced in the future'.[46] But, unlike Biosphere 2,
this experiment has been unintentional. The nineteenth-
century industrialists who powered the modern world
by burning coal, oil and gas on a massive scale never
imagined that one of the consequences of their industry
would be to warm the planet. The experiment identified
by Revelle and Suess in the late 1950s was recognised
30 years later, when British Prime Minister Margaret
Thatcher delivered the first speech on the topic of
climate change by a major political figure. On 27
September 1988, addressing the annual dinner of the
Royal Society in London, Thatcher echoed the words

of Revelle and Suess. She warned of 'a global heat trap which could lead to climatic instability' and referred to the possibility that 'we have unwittingly begun a massive experiment with the system of this planet itself'.[47]

Revelle and Suess's 'large-scale geophysical experiment' of a greenhouse gas-warmed planet was unplanned in the sense that climate change was an inadvertent byproduct of securing a major public good. This good was cheap, reliable, versatile and abundant energy. In contrast, the deliberate injection of aerosols into the stratosphere is an intentional experiment where the resulting engineered climate *is* the public good! Unlike in the case of combusting fossil fuels, here there is no objective to the technology other than to create a global climate less threatening than would otherwise be the case. The public good becomes the 'undoing', or in some way the 'offsetting', of the unintentional consequences of fossil-fuel use.

This creates a very different set of evaluative procedures as to whether or not stratospheric aerosol injection should proceed. It places a specific responsibility on those who advocate the technology to foresee, and then to estimate, the possible unintended consequences of implementation. The risk calculus with regards to greenhouse gas emissions requires the welfare benefits of the continuing use of cheap and reliable fossil energy – these benefits are quite well known – to be weighed against the consequences of an altered climate – these are tricky to isolate and include benefits as well as costs. The risk calculus for stratospheric aerosol injection is quite different. The welfare benefit of reducing the build-up of heat in the global atmosphere and ocean – a benefit that is poorly understood – has to be weighed against a range of possible risks and unintended consequences, most of which are incalculable.

As I have shown, the distinction between empirical research into aerosol injection technology and its implementation breaks down when questions are raised about regional benefits and risks. There is no miniature Earth upon which to test this; Biosphere 2 will not do. The natural analogues of volcanic eruptions can be suggestive of some of the risks, but never definitive. And physical theory of climate and ecosystem feedbacks is crude. Computer simulation models, even operating at the limits of calculative power, are therefore far from accurate or precise enough to determine what these risks might be. At the very most such models offer a range of conditional probabilities – often conflicting – for some large-scale climate outcomes.

The list that follows is limited to outlining just some of the possible *environmental* consequences of stratospheric aerosol injection deployment. It excludes the various social, economic, political or military repercussions that might follow intervention in the stratosphere, some of which I examined in the previous chapter.

- *Untoward effects on regional climate* This is perhaps the greatest concern of all. As I showed in Chapter 2, controlling global temperature is not the same thing as controlling regional climate and local weather. Following a global aerosol injection programme, the potential for major regional disturbances to climate would be great – particularly in the form of rainfall and storms. Since not all local climates can be simultaneously optimised, this may well lead to compensating local injection programmes being pursued by regional or national authorities.

- *Stratospheric ozone depletion* Aerosol particles offer surfaces for chemical reactions in the stratosphere that destroy ozone. One consequence of injection – at least while ozone-depleting chemicals remain in production – would therefore be enhanced rates of ozone depletion and the attendant increase in melanoma cancers.

- *Effects on plants* Loading the stratosphere with aerosols would change the proportions of direct and diffuse solar radiation received at the Earth's surface. The effects of such changes on plant growth and plant health are poorly known.
- *More acid deposition* As the sulphate aerosols slowly sediment through the Earth's atmosphere, they would contribute to lower tropospheric acidity, thus adding to ecosystem and human health burdens.
- *Whitening the sky* Aerosols in the stratosphere would likely give a white, cloudy appearance to the sky. As in some Chinese cities today, blue skies would become the exception. On the other hand, sunsets might become more vivid, changing the aesthetics of the sky.
- *Less sunlight for solar power* Screening out the sun's heat through an aerosol layer would reduce the renewable resource available for solar power systems, solar radiation – one of the favoured energy sources alternative to fossil fuels.
- *Environmental impacts of implementation* Whichever delivery system were to be adopted for the aerosol injection – planes, missiles, balloon-and-hose – there would be a range of environmental

side-effects from the large-scale deployment of such systems.

- *The climatic consequences of cessation* Were aerosol injection to cease after several years or decades, whether on plan or not, a significant jolt to the climate system would be likely to follow, as the heating effect of accumulated greenhouse gases would suddenly be unmasked. Such a jolt would cause an even more rapid change in climate than would have occurred without aerosol injection.

- *Changes to environmental systems as yet not known* The environmental consequences listed above have either been envisaged or else simulated in models. In a complex non-linear system such as the Earth there will be other consequences of aerosol injection, which are not yet foreseen. And of course carbon dioxide would continue to accumulate in the atmosphere and in the ocean, leading to further ocean acidification.

For these environmental reasons, combined with the even less foreseeable social and political repercussions, the result of deploying aerosol injection technology is that humans would become experimental subjects. The unintended outcomes of creating a thermostat in the sky would place an almost unbearable responsibility upon those who would finally take the decision to implement. This is the 'big red button' metaphor of the age of nuclear weapon arsenals. For many, the sheer range, extent and legacy of the attendant risks associated with stratospheric aerosol injection prohibits unleashing the experiment.

Plan B and infinite regress

In his introduction to a recent special issue of the journal *Environmental Values*, which dealt with synthetic biology, ethicist John O'Neill reflects on the practice of 'engineering'.[48] Drawing upon practices of genetic, biotech and social engineering, O'Neill asks whether the widespread popular antipathy for engineering in the biological and social spheres is fully warranted. He challenges the pejorative view that the engineering mind is inspired by a (dangerous) desire for controlling a system on the basis of knowledgeable outcomes. O'Neill offers a different reading of the engineer, as one who is always precautionary, always thinking of what may go wrong and trying to design in suitable safeguards or responses.

But the problem for engineers who promote aerosol injection technology is that there are just *so* many things that could possibly go wrong with a system that is global in scale. Even if O'Neill is right about the engineers' instinct to be precautionary, they still cannot foresee the repercussions of the Plan B of climate engineering. The precautious engineering mind that designs this particular thermostat is faced with irreducible uncertainties. There are *prima facie* limits to acquiring the knowledge that is necessary if one is to foresee all consequences. The economist Stephen Marglin has put the problem succinctly:

> If the only certainty about the future is that the future is uncertain, if the only sure thing is that we are in for surprises, then no amount of planning, no amount of prescription, can deal with the contingencies that the future will reveal.[49]

As the ramifications of the thermostat for regional climate and local weather emerge *ex post*, the spectre of international lawsuits for damage caused will arise (see Box 4.1); and there will be demands for further tinkering with the global atmosphere. Implementing Plan B will in due course necessitate further interventions in the skies. Plan C will then be promoted – perhaps adding aerosols asymmetrically between hemispheres, to compensate for some claimed malfunction of the Brazilian rain system. And so an infinite regress of 'plans' will occur – Plans D, E, F and so on – with layer upon layer of human intervention inscribed in the skies. Experimentation with the world's climate will never end. No wonder that artists are intrigued by this new creative heavenly canvas that the geoengineers are opening for experimentation.

Box 4.1 Who Gets the Blame for Bad Weather?

It is not easy to find out who or what is to blame for bad weather, even less so when the climate is changing. In some cultures the spirits or the gods are to blame, agents who sometimes are provoked to such retribution through wilful human misbehaviour or violations of important social or moral norms. In many western societies today, 'Acts of God' are still invoked in certain insurance claims against bad weather. And then there is blame that is attached to negligent human authorities for turning a weather extreme into a major public disaster: poor planning, sloppy design, inadequate warnings, and so on.

Can science help by attributing specific weather extremes to particular physical causes, whether these be natural – such as sunspots – or human-related – such as greenhouse gas emissions? Some climate scientists have been developing methods based on computer simulation models in order to attribute specific weather extremes to human causes, notably elevated concentrations of atmospheric greenhouse gases. This has led to the concept of fractional attributable risk, according to which the odds of a specific weather event's occurring in a specific place is attributable to a given causal influence. Thus it is claimed that the 2003 European summer heatwave was substantially attributable to elevated greenhouse concentrations, whereas the heavy rainfall in Thailand causing extensive flooding in the summer of 2011 was natural in origin. In the work done so far, probabilistic event attribution creates two categories of weather extreme, and therefore two categories of social and ecological impact: 'human-caused weather' (and impact) and 'tough-luck weather' (and impact).

These scientists argue that there is an urgent need to develop further this science of weather event attribution to assist with decisions about the allocation of international adaptation funds. One of them claims that,

> because [adaptation] money is on the table, it is suddenly going to be in everyone's interest to be a victim of [anthropogenic] climate change . . . so we need urgently to develop the science base to be able to distinguish genuine impacts of climate change from unfortunate consequences of bad weather.[50]

The implication of this logic is that compensation might be available for the victims of the former kind of bad weather, but not for those of the latter.

I can easily see how an extension of such attribution analysis would be called upon in the case of stratospheric aerosol injection. Following implementation of the thermostat, many interested parties will ask the question: Was this particular extreme typhoon caused by artificially injected aerosols, by elevated concentrations of greenhouse gases or was it a naturally occurring tropical storm? Pressing liability claims for the social or ecological impacts of non-natural weather extremes will quickly follow, as was foreseen by the Scottish citizen Thomas Crowley quoted in Chapter 3. Yet the simulation studies behind weather event attribution are deficient in many respects. Isolating the effects of different causal agents is very difficult. Artificial aerosols, sunspots, carbon dioxide from power stations, methane from cattle herds – all interact in complex, non-linear ways, which models cannot reliably capture. Different models yield very different results. And, in any case, extreme weather is always mediated by social, institutional and political factors that have nothing to do with meteorology.

Rather than making it possible to identify the causes behind weather extremes in a thermostat-controlled experimental climate, weather event attribution is likely to aggravate the political and ethical conflicts that aerosol injection technology will have unleashed.

This is a manifestation of Ulrich Beck's risk society *in extremis*: the modern human lives in a perpetual state of fear and danger, so Beck claims, anxious about the unseen pathological consequences of ubiquitous and mundane technologies. These hidden risks connect together previously separated peoples and create a more ominous form of cosmopolitanism than that idealised by Immanuel Kant. In the twenty-first century it is the uneven distribution of risks that drives politics – as much as does the unequal distribution of wealth and power.

And yet, if we listen to the advocates of stratospheric aerosol injection, the response to such technologically induced climatic anxiety is to initiate more of the same. If anthropogenic climate change is the condition against which we react, then another dose of anthropogenic climate change – this time brought about deliberately, through the creation of a thermostat in the sky – is surely a vain hope. All of the claimed dangers of greenhouse gas-induced climate change – disturbances to ecosystems, untoward changes in local weather, uneven distribution of winners and losers and so on – are to be reconfigured through another round of climatic manipulation. This technical impulse is now to be applied to the planet itself – and by this action not just the material Earth is to be changed, but also how humans designate themselves as exceptional.

Re-making the earth, re-making the human

The above analysis of the risks of stratospheric aerosol injection shows that, once one is embarked upon it, there is no end to the experimentation. Aerosol injection is not a technology that can undo changes to the climate once these have already been induced; it cannot intervene mechanically in a system to restore some previous, idealised condition of nature. The metaphor of restoration does not work here. If stratospheric aerosol injection is to be implemented, the Earth would be remade in a way and on a scale that humans have never before accomplished or witnessed. The climate would become artificial in the literal sense of becoming an artefact – a product of human endeavour. We could appropriate Bill McKibbin's neologism 'Eaarth' to describe such a planet – one still recognisable and yet fundamentally different.

What would it be like to live in this brave new world? What would it be like to be subject to this perpetual global experiment? We do not know, but we can at least try to imagine. And so artists, poets and novelists have undertaken some of the difficult imaginative work. In his exhibition project 'Hacking the Future and Planet', the Austrian artist Klaus Schlafer brought together well-known artists and art critics to explore the boundaries between scientific and artistic experimentation around weather making. One of his public art projects was to whitewash the main square in the town of Pischeldorf, as an artistic demonstration of a sunlight reflection method. Through such sculpted artefacts, Schlafer wants to intervene dramatically in the world and to provoke public conversations about the desirability of remaking the climate. His 'cooling laboratory' – accompanied by films, lectures, exhibits and a book – has toured Austrian cultural festivals to this end.

A different example of transcending science in order to sensitise the human imagination to climate engineering might be found in the speculative fiction of the novelist Margaret Atwood. Her 2013 book *Madd-Addam* concluded a dystopian vision of a future where most of humanity has been wiped out. Her trilogy, comprising *Oryx and Crake* and *The Year of the Flood*, reflects a concern about the consequences of human engineering of the material world. Although focused on bio-engineering and on the emergence of a quasi-human species, Atwood creates a powerful narrative about experimentality gone wrong. It is not too difficult to see it applying to a dysfunctional world in which battles over the stratospheric thermostat create perpetually anarchic climates.

Going beyond artists, it is also necessary to ask citizens what they think about such climate experimentation. In a carefully conducted investigation into the reactions of different British social groups to the idea of sunlight reflection, Phil Macnaghten and Bron Szerszynski reported the following conversation among a focus group of women aged between 25 and 40, all mothers of young children and involved in their local community:

KAITLYN   The thing that scares me is that the experiment will be done in our [life times] . . .

NATALIE   It's such a huge thing, though, isn't it?

KAITLYN   Yeah. The experiment will be while we're here . . . and . . . for our children. What if the experiment goes wrong? Then what happens?

LILLIAN   Do you think it could destroy the Earth?

KAITLYN   Yeah, it could go the other way. How can you test . . .? Can it be tested in a laboratory? But then it's got to go out there.

MODERATOR   Yeah, sure. So that's the big question for you. We'll be living the experiment, in a sense. Is that what you're saying?

KAITLYN   Yeah.

NATALIE   We're the lab rats.[51]

The prospect of engineering the climate through aerosol injection is not just about imagining the changed physical conditions of climate and weather on the planet. It is not just about reconfiguring the Earth as Eaarth. It is also about imagining how such planetary experimentation would change how humans think of themselves. The realisation that humans can manufacture the climate through a global thermostat would change our self-image in fundamental ways. Some of us may discover ourselves to be Natalie's 'lab rats', powerless to escape from conditions we neither expressly sought nor mandated. And, as has been suggested by Bron Szerszynski, the label *Homo faber* – literally the 'making human' – might be more appropriate for us than *Homo sapiens* – the 'wise human'. To follow McKibbin's terminology, the brave new world being brought into existence would comprise not Earthly humans, but Eaarthly huumans.

Deploying stratospheric aerosol injection to control the world's climate would further blur the already fading boundary we have constructed between the human and the natural. Just as synthetic biology will create new quasi-human species – as in Atwood's trilogy

– so will stratospheric aerosol injection create new quasi-human climates. I concede that some will argue that this blurring is both inevitable and desirable. After all, these two categories – the human and the natural – are artificially constructed and should be done away with. Why should 'messing with nature' present us with a problem?

But not everything that seems inevitable is good or desirable. Not everything we imagine *could* happen is something we *want* to happen. 'Because we can' is never an adequate justification for human beings, for whom the moral categories of wise and foolish, good and bad, right and wrong are unavoidable – even if we do not always know where to locate them. In the end, the boundaries by which we navigate and make sense of the world have to be chosen. (And as I explain in Chapter 5, we also have to choose our problems wisely.) So I believe there remains some value in retaining a distinction between inadvertent and intentional experimentation. The American philosopher Christopher Preston describes the psychological cost of deliberate aerosol engineering this way:

> At every moment it would be our responsibility to ensure that climate was hospitable. Rather than viewing nature in the traditional fashion as a deep source of

> solace and meaning, we might start to view climate as a constant (and self-created) threat, leading to existential anxiety [that] would plague us.[52]

I do not believe that human beings are ready for a global thermostat and the new responsibilities it brings, just as they are not ready for human cloning or germ-line modification. This is not because some transcendent mandate not to mess with a sacred nature would be violated. My argument is simply that the environmental, political and psychological costs of designing global climate through aerosol injections seem overwhelmingly to outweigh any assumed benefits.

Summary

The key difference between the two experimental conditions described above revolves around intentionality. Through the transformation of land, the production of energy, the consumption of materials, and the begetting of children humans have altered the physical processes of climate around the world. And we have been doing so for a very long time, certainly for a thousand years, maybe for ten thousand. But the experimentation with climate that has resulted from these activities has been inadvertent and unintentional. These activities have been undertaken for reasons completely unrelated to the climate, the primary one being that (for most) they have brought significant welfare benefits and (for some) significant profits.

But to deliberately change the condition of the planet's atmosphere in pursuit of the 'public good', in order to compensate for an induced planetary heating, is an entirely different form of experimentation. It suggests a supreme confidence in human knowledge and ingenuity – a confidence approaching arrogance. Even to research the technologies for doing so reveals a certain poverty of the imagination, a preference for technical calculus that has little regard for the relational, creative and spiritual dimensions of *anthropos*, the human being. This is why interventions from those beyond the geoclique are needed to bring us to our senses: anthropologists, artists, historians, philosophers, poets – analysts and storytellers who are attuned to the paradoxes of the human condition.

Some experiments are unavoidable; and some experiments are worth conducting. But all experiments that fall into these categories are limited in scale and scope. And experiments are entered upon because the outcome is not known. Will patients recover after receiving a new drug? Do crops grow faster after the application of a new fertiliser? Will this new material conduct electricity efficiently? Absurdly, the experiment with stratospheric aerosol injection would have to be justified on the (false) grounds that the outcome is already known, that it will deliver the desired 'beneficial' outcome, that a global thermostat *can* stabilise the climate.

On the contrary. To embark on this course of action would indeed be to conduct a giant experiment, to take a leap in the dark. It is not possible to know what the consequences of such engineering would be. No matter how much we are attracted to machinic metaphors in describing the natural world – picturing systems that can be re-tuned or re-engineered according to human desire or need – I do not believe the human mind has the ability to fathom the intricacies of how the planet functions. The simulation models upon which aerosol injection technology would rely are like calculative cartoons when it comes to making long-term predictions. There are limits to human knowledge; our species is a product of evolution, not its author or controller.

To proceed with stratospheric aerosol injection is to experiment with the Earth. It is to remake its atmosphere, it is to fabricate its climate. We fool ourselves into thinking that climate can be restored to some pre-industrial condition, or even that it can be reconfigured into some more benign state. And the experiment will be a never ending and open-ended one. It will place on humans new responsibilities requiring moral or ethical capacities for the possession of which there is little evidence. Stratospheric aerosol injection *is* an avoidable experiment; it is not one that we *should* conduct.

Thus far I have presented a series of arguments pointing out why to regulate global climate by using solar intervention technologies in the stratosphere is undesirable, ungovernable and unreliable. Science cannot fix climate change this way. In the final chapter I offer a different perspective on the unstable relationship between humans and their climate. This reframing of the problem suggests a different role for science and technology from the one envisaged by those who contemplate a global thermostat in the sky.

FIVE

Reframing the (Climate) Problem

Approach the goal obliquely

Winston Churchill was a man of many parts. He was a prodigious writer and a Nobel Laureate for Literature; he was the embodiment of British aristocratic heritage; and he was a great political opportunist. He was also a successful war leader and regarded himself as a creative and visionary military strategist. While some of his military ideas were impractical and far-fetched, one consistent belief he held across both World Wars was about the importance of the Mediterranean as a theatre of war for the defeat of Germany. Churchill believed that attacking the 'soft underbelly of the Axis' through the Mediterranean was a strategy that would yield greater military and political gains at lesser cost of human life than full frontal assaults on German power in northern and western Europe. In 1915 his plan was to force the Dardanelles to trigger the collapse of Turkey, offer direct support to Russia, and hence break the stalemate on the Western Front. In 1943 his plan was to attack Germany through Italy and the Balkans in preference to a full frontal attack on Hitler's Atlantic Wall.

The precise details of Churchill's Mediterranean strategy are not what interests me here. Rather it is his broader thinking that I think is of relevance for developing strategies for tackling the risks and dangers of climate change. Churchill's belief in the Mediterranean route to victory illustrates an important principle: sometimes the most effective way of reaching one's goal is to approach it obliquely rather than to take what appears to be the most direct route.

A different illustration of the same principle comes from the innovative landscape designs of Lancelot 'Capability' Brown in eighteenth-century England.[53] Brown designed the estate landscapes for over a hundred English stately homes, and his distinctive contribution to landscape design is captured in the phrase 'lose the object and draw nigh obliquely'. His landscapes would offer the visitor an initial glimpse of the final destination – the stately home – but only briefly. The driveway then took the visitor along a winding pathway of graceful curves, through delightful woodland glades and meadows, over dammed streams and lakes, before delivering him or her at the door of the grand home at the heart of the estate.

How do these two examples act as analogies for thinking strategically about climate change? They do so by suggesting that there may be more effective ways of achieving the goal of minimising the risks of a change in climate than to use brute-force methods to try to stop this change from happening. Creating a global thermostat through stratospheric aerosol injection (Plan B) launches a full-frontal technological offensive against climate change, much as the comprehensive multilateralism of the United Nations' targets and timetables approach to emissions reduction (Plan A) offers a full-frontal geopolitical assault. But perhaps there are different ways of understanding what is problematic about climate change, framings that suggest different, less direct ways of attending to the issues raised. And by reframing what the problems are, perhaps there are better ways for science and technology to contribute to human welfare than offering to construct a thermostat in the sky.

It is important to choose one's problems carefully. And how a problem is framed opens up or forecloses certain types of solutions. Problem framing is therefore always a political act requiring insight, pragmatism and wise judgement. Successful politicians – like successful generals – always seem to know which battles to fight. In this chapter I challenge the conventional framing of global warming as being 'the problem' that must be fixed. This framing obscures a more diverse range of welfare goals – sustainable energy, human health, food security, ecosystem integrity – for which there is no silver bullet and for which a global thermostat is mostly irrelevant. I argue that climate change is a wicked problem, one that is not solvable through a universal strategy – whether technology in the skies or multilateralism on the ground. By analogy with my earlier examples, the task is to find the 'soft underbelly' of climate change, to find oblique policies that advance human welfare while at the same time delivering outcomes consistent with the goal of minimising climatic risks. This approach can be called climate pragmatism; it is inspired by scientific and political pluralism and puts human values before Promethean technologies.

Climate change is wicked[54]

The advocates of researching into stratospheric aerosol injection argue that the technology may be needed in the case of a climate emergency or, at the very least, that it may buy time to secure an energy transition.

Lord Rees, the Astronomer Royal and former president of the Royal Society, is just one of many voices – echoing the geoclique – that have recently made this precise argument. But to research aerosol injection is to contemplate the possibility that it may be deployed. And the narrative for contemplating the global thermostat, which I have endeavoured to confront in this book, requires a justification, a goal, and a means to achieve the goal. The justification is the threat of a global climate emergency and the crossing of tipping points – putative dangers that, it is claimed, stratospheric aerosol injection would avert. The goal in this narrative is to maintain the red line of no more than two degrees of warming – a goal that, it is claimed, stratospheric aerosol injection could secure. And the means of controlling the thermostat are Earth system models that, it is claimed, would guide the geoengineers in their task.

I have argued against this narrative and its presumption that science can 'fix' climate change. I have demonstrated why the idea of a global thermostat in the stratosphere is undesirable (controlling global temperature does nothing towards controlling local weather), ungovernable (there is no prospect for a universally agreed thermostat setting) and unattainable (the unintended consequences of aerosol injection would lead to unending experimentation). My argument is not that science and technology have no role in reducing the risks associated with climate change. Nor is it that implementing aerosol injection would constitute a sacrilege against an inviolable nature. My argument is simply that the attendant risks of implementing a thermostat – as far as they can be imagined and quantified – greatly outweigh any plausible benefits that might result. Another way of expressing my position would be to say that stratospheric aerosol injection is the wrong sort of solution to the wrong sort of problem. Human-induced climate change is not the sort of problem that lends itself to technological end-of-pipe solutions. It is not like asbestos in buildings – a hazard for which there is a technical solution. Climate change is a 'wicked problem' and needs to be approached differently, obliquely, if its dangers are going to be defused.

More than forty years ago planning theorist Horst Rittel proposed the phrase 'wicked problems' to describe a category of public policy concerns that defied rational and optimal solutions. Wicked problems have no definitive formulation and can be considered a symptom of yet other interrelated problems. Wicked problems frequently emerge from unbounded, complex and imperfectly understood systems. Their solution is beyond the reach of mere technical knowledge and traditional forms of governance. Solutions to wicked problems are impossible to effect because of complex system interdependencies: a solution to one facet of a wicked problem will often reveal or create other, even more intractable problems demanding further intervention. Climate change possesses all the attributes of a wicked problem.

'Tame' problems, on the other hand, while they may be technically complicated, have relatively well-defined and achievable end states with a limited number of system interactions. They are potentially solvable. The depletion of stratospheric ozone would fall into the category of a relatively tame problem.

Failure to understand and treat climate change as a wicked problem has led to the search for global solutions that possess elements that are inadequate, inappropriate or obstructive. Adopting two degrees of global warming as the singular overarching goal of international climate policy is to misread the nature of the problem. It assumes that climate change is a tame problem rather than a wicked one. It has led to Plan A – with its vain hopes that market-based emissions trading and consumer-oriented social marketing would deliver the emissions reductions calculated to be necessary. And this failure has in turn led to the contemplation of Plan B – a thermostat in the sky. These solutions rely too heavily on regime theory in politics, rational choice theory in economics and social coercion in human behaviour management (Plan A) and on geophysical control engineering for the implementation of stratospheric aerosol injection (Plan B). Plans A and B both underestimate the wickedness of climate change and overestimate the abilities of political, economic, psychological and scientific knowledge to control the changing climate.

The conventional framing of climate change with its two-degree policy goal begs a more fundamental question. What is the ultimate performance metric for the human species, what is it that 'we' are seeking to optimise? Is restabilising global climate the project we should put before all others? Why should it not be to stabilise world population, or to make poverty history? And what about maximising gross domestic product, increasing the sum of global happiness or securing political emancipation for all? Or is the ultimate performance metric for the human species simply to survive? Setting out on a global project to restabilise the climate as though this were the ultimate public good hardly does justice to the range and diversity of legitimate human aspirations. It certainly is not the self-evident universal goal that a global thermostat would require as its justification.

Wicked problems do not yield to brute-force technological solutions such as the thermostat. Stratospheric aerosol injection is a flawed idea: one that seeks an illusory solution to the wrong problem. Used in this way, science cannot fix climate change. Wicked problems need framings that allow a proliferation of diverse solutions, since these solutions should attend to different problem elements. They should appeal to plural, even contradictory, human values. My argument is that, when solutions to previously framed problems are no longer forthcoming or lead to dangerous adventurism, we should be willing to restructure the problem. This is what is meant by social learning. In the section below I offer a different way of framing the challenges that are posed by climate change; it is called climate pragmatism. I suggest that, by reframing the problems of a changing climate, science and technology make contributions to human welfare that are very different from serving the Promethean vision of a techno-fix for climate change.

Climate pragmatism

All things in the world are materially and politically related, the more so as connectivities proliferate and their reaction times accelerate. Globalisation is a social phenomenon that emphasises our material interdependency. The subprime mortgage lenders in the US Midwest 'cause' economic collapse in Greece just as much as the flapping of a butterfly's wings in Brazil 'causes' a cyclone in Bangladesh. Amid this hyperconnectivity, the task of wise problem framing is to choose judiciously where to cleave this seamless reality. Out of the multitude of problems that it is *possible* to define and tackle, which are the ones that are *wisest* to define and tackle? Such decisions are inevitably political and ethical in nature.

Over recent years an eclectic group of scholars and analysts known as the Hartwell Group has developed the idea of climate pragmatism in response to this challenge.[55] Climate pragmatism is based on two principles that undermine the conventional framing of human-induced climate change, which became solidified during the early 1990s. The first principle is to decouple the energy question – that is, the question of how to meet the growing demand for energy reliably, cheaply and sustainably – from the climate question. Yes, there are interdependencies between the ways in which these two questions get answered, but climate pragmatism argues that it makes better sense to tackle them as distinct problems. The second principle is to recognise the many different ways in which human activities alter the composition and functioning of the atmosphere – and that each of them produces different welfare risks and hazards at different scales. Emissions from fossil-fuel combustion, from land use change, from industrial manufacturing – they all alter the climate system, but in different ways, at different space and time scales and with different effects. Grouping these emissions together into one basket – of pollutants that are indexed for policy purposes solely by their contribution to global warming and for which a single solution is sought – is an example of unwise problem framing and solution seeking.

These two principles of climate pragmatism turn the singular problem of climate change – which science is seeking to fix through stratospheric aerosol injection – into a tripartite problem: reducing weather risks; improving air quality; innovating in the search for cheap, reliable, clean energy. And the global thermostat will resolve none of them. In contrast to the universal policy goal of limiting global warming to no more than two degrees (a goal that has driven climate politics for most of the past 25 years), climate pragmatism opens up a portfolio of policy goals. It is a portfolio for which a diverse range of strategies, instruments and interest coalitions is appropriate. Returning to my analogies with Winston Churchill and Capability Brown, one might say that a more pragmatic approach to tackling climate change offers three tasks by which the goal of reducing the risks and dangers associated with a changing atmosphere and climate may – indirectly, but more plausibly – be reached. Let me briefly summarise them.

The first task would be to enhance social resilience to meteorological extremes and thereby reduce the magnitude of weather risks. This is the true public good: well-adapted societies, and not the illusory idea of a stable global climate. Isolating the precise causes of meteorological extremes is of secondary importance. It will never be possible, in specific cases, to apportion them with any precision, considering the full range of human and non-human influences – sun spots, carbon dioxide, volcanoes, dust, methane, black soot. And, in any case, weather event attribution is not the same as disaster attribution. Weather risks are as much a function of social, institutional and engineering processes as they are a result of meteorology. Since all societies are maladjusted to climate to varying degrees, weather extremes and climate variability impose (uneven) costs on all societies – as well as generating certain benefits, of course.

Climate pragmatism therefore seeks to evolve new technologies, institutions and management practices – to harness the creativity of science – so as to minimise the costs and damages wrought by weather and climate. This is not a universal, one-size-fits-all solution, but it requires different types of investments in different regions and at different scales. The World Meteorological Organisation's recent initiative for a Global Framework for Climate Services is one example of such climate risk management strategies. Others are coastal and riverine ecological engineering, new micro-insurance initiatives, community-based adaptation, improved early warning systems. As I have shown in Chapters 2 and 4, even if a thermostat were successful in rebalancing radiation flows at a global scale, it would inevitably further destabilise regional climate and weather. Even if global temperature is regulated to be only two degrees warmer than it was in the nineteenth century, the need for improved social resilience and greater adaptive capacity will remain.

The second task of climate pragmatism is to reduce emissions of atmospheric pollutants other than carbon dioxide that have environmental and health costs. These would include black soot, methane, stratospheric ozone-depleting halocarbon chemicals, mercury and

precursors to tropospheric ozone. Most of these pollutants have climatic effects – some regional, some
global. Taken as a group, their contribution to global
warming is almost as large as that of carbon dioxide
released from fossil-fuel burning. These pollutants also
diminish human and ecosystem health. A pragmatic
approach to climate policy would therefore be to
develop and harness new and more efficient technologies for the reduction of these health-harming and
climate-altering emissions. Such technologies would be
regulated and incentivised through sectoral, regional or
bi-lateral agreements. This would stand in contrast to
the universal global regime favoured by the United
Nations Framework Convention on Climate Change
(UNFCCC), in which the global-warming properties
of pollutants are used as the sole index by which to
trade them along with carbon dioxide in managed
markets.

Climate pragmatism applied in this way is a good
example of the oblique approach to tackling climate
change: regulating non-carbon dioxide emissions brings
significant public health benefits, and by so doing it also
reduces human influence on the climate system. It is a
case of promoting welfare policies that have climate
co-benefits rather than pursuing climate policies that
have welfare co-benefits.

The third task advocated by climate pragmatists is to meet the growing demand for energy in the world cheaply, reliably and sustainably over the decades to come. Global energy use is expected to increase by nearly 50 per cent by 2035, due both to population growth and to the needs of human development and poverty alleviation. This will require a transition away from fossil fuels towards partly or fully renewable technologies – and therefore a reigning in of carbon dioxide emissions from the energy sector. But this will not be achieved quickly. And it won't be achieved without large-scale innovation – researching, developing, demonstrating and deploying new energy technologies that can compete with conventional and unconventional coal, oil and gas. Wind turbines and solar tiles on roofs will not do; nor will carbon-trading schemes provide sufficient incentives for the scale of investment required.

This challenge requires political commitment to innovation policies and to delivering massive new investments in energy science and in energy technology development and deployment. Such investments can be achieved through unilateral, bilateral or minilateral initiatives; they do not require a global treaty, signed by 193 countries and tied to nominal timelines dictated by the still poorly understood global carbon cycle. One way to secure the needed investment – a way advocated by climate pragmatists – is through a hypothecated carbon tax that would initially be very low, before it would gradually rise. This is a long-term transition that would build slowly. The goal of this innovation policy is to make renewable energy cheap, not to make fossil-fuel energy expensive. A related and appropriate goal of science and technology in this area is to research and develop benign technologies for removing carbon dioxide from the atmosphere; such technologies are sometimes called 'negative emissions' technologies or carbon dioxide removal technologies.

If the objectives of climate policy are reorganised around these diverse and pragmatic welfare goals, then the justification for a global thermostat to control global temperature is made obsolete. The thermostat is, at best, irrelevant for securing these goals and, at worst, it could make them harder to attain. The thermostat would do nothing to make society more resilient to weather risks. Indeed, as I have shown, stratospheric aerosol injection would further destabilise regional climate systems that generate meteorological and climatic extremes. Nor would stratospheric aerosol injection do anything to improve air quality by reducing the atmospheric pollutants that result from human activities. The health and environmental evils caused by such emissions would continue. And establishing a thermostat in the sky would do nothing to contain the volume of carbon dioxide that enters the global atmosphere. Injecting aerosols into the stratosphere would do nothing to incentivise the diversification of cheap, reliable and sustainable energy sources. Nor would it slow down the acidification of the world's oceans.

I will say it again: stratospheric aerosol injection is a flawed idea – an idea that seeks an illusory solution to the wrong problem.

Why science cannot fix climate change

How humans experience weather and climate risks, and the extent to which human activities are altering such risks, are, both, matters deeply bound up with the material and cultural forms of life we have inherited from the past and propagate into the future. The dangers of climate change are not somehow 'out there', like a danger of alien invasion – external to the way we live and organise ourselves. The dangers of climate change are 'in here' – a function of human technologies, social relationships, economic and political systems.

This places a limit on what science can achieve (see Box 5.1). Yes, the risks associated with a changing climate are mediated by the physical processes of the atmosphere, oceans, ice sheets and biosphere. And science is well placed to help us understand how these processes work. But the climate risks we imagine or experience are also powerfully mediated by the ways we live and argue together. This means that to claim to be able to 'fix' climate change – in the sense of reducing or eliminating weather risks and the associated dangers of climate change – would be to claim to 'fix' society in some way. Any purely technological claim to fix climate change, such as the suggested global thermostat, misses at least half the story.

Box 5.1 Technologies of Humility

In his famous Rede lecture 'The Two Cultures', delivered at the University of Cambridge in 1959, civil servant and novelist C. P. Snow offered an optimistic vision of the future. The onward and upward advance of science and technology would modernise the world, eradicate poverty and close the gap between rich and poor. In the 55 years that have passed since then, technology *has* changed every corner of the world and has contributed to a general increase in life expectancy in most countries. But Snow's optimism that scientific discovery would secure continued improvement in the well-being of humanity has proved misplaced. It is not that Snow oversold the transformative power of science and technology and its potential for improving the quality of life. It's more that he didn't sufficiently appreciate that science, just by itself, will only ever offer a partial cure to the afflictions of mankind.

I have argued in this book that, by developing and deploying stratospheric aerosol injection technology, science cannot 'fix' climate change. Indeed I have argued that, by attempting to so implement this technology, science will probably add further suffering, antagonism and conflict to the world. Similarly, science cannot 'fix' the problems that preoccupied Snow in the late 1950s and still scourge the world today: hunger, poverty, unequal life chances, morbidity, inter-communal violence. This is because these miseries emerge out of specific social practices, cultural identities and patterns of human relationship, in sum, they result from the exercise of power by the few over the many or the many over the few. Attending to the pathologies that pervade these dimensions of human conduct requires political, ethical and social interventions and regulation *before* science could 'fix' them.

Such interventions are what Sheila Jasanoff, professor of science studies at Harvard University, has called 'technologies of humility': attitudes and habits that recognise the limits of human knowledge, the complexity of socio-technical systems and the primacy of moral considerations in governing human actions and behaviours. To these humble dispositions she counterposes the 'technologies of hubris' – beliefs and convictions that it is through advances in science and technology that human problems will be solved. The latter were on display 55 years ago, as C. P. Snow's claim reveals:

> There is no getting away from it. It is technically possible to carry out the scientific revolution in India, Africa, South-east Asia, Latin America, the Middle-East, within fifty years. There is no excuse . . . not to know that *this is the one way out* through the three menaces which stand in our way – H-bomb war, over-population, the gap between rich and poor.[56]

Science is concerned with what is knowable through material observation and experiment, and technologies convert these 'knowable' facts into material artefacts that alter the functioning of physical matter and systems. Conceptual models of scientific prediction and engineering control continue to dominate the discourse of improvement, not least in the arguments in favour of a thermostat in the sky. But the 'can do' attitude of the geoclique must give way to the 'should we' questions raised through ethical, moral and political reflection. The irony is that the worldwide expansion of Snow's valorised culture of science and technology – of which the global thermostat is the progeny – has simply shown the enduring necessity of his 'other' culture – philosophy, religion, ethics and the humanities.

And this is what I object to most strongly about the call to research, and prospectively to deploy, stratospheric aerosol injection technologies and other sunlight reflection methods that seek to regulate global temperature. The justificatory narrative for a thermostat focuses on a single geophysical control variable – global temperature or the global heat budget – as the ultimate object of climate policy. It claims that such intervention may be warranted on the basis of a climate emergency; and it suggests the possibility of exercising technocontrol over the weather in the name of a noble cause. It claims that a stable global temperature is a public good for which we should strive. By doing so, the geoclique and its acolytes seek, ironically, to naturalise human-induced climate change – to place it beyond the domain of politics. Even raising the idea of a global climate emergency that may 'demand' the insertion of a planetary thermostat elides the political and the ethical. Climate emergencies are made, not discovered, and what matters most is who announces them and for what purpose.

One of the (many) dividing lines in debates about climate change is between those who think that climate change is such a totalising concern that normal politics must be suspended and those who recognise that human life is political before it is material. The iconoclastic scientist James Lovelock clearly sides with the former. A few years ago he claimed that, in the light of human-induced climate change, we need a more authoritarian world:

> What's the alternative to democracy? There isn't one. But even the best democracies agree that when a major war approaches, democracy must be put on hold for the time being. I have a feeling that climate change may be an issue as severe as a war. It may be necessary to put democracy on hold for a while.[57]

Seeking to manipulate global climate through the use of sunlight reflection methods is nothing short of an ideological project of techno-mastery, justified in the name of science by being declared to be a necessary response to a global climate emergency. It is a project that indeed runs the risk of putting democracy on hold – or of presenting it with a challenge to which democracy is unable to respond, which is equally concerning.

But the argument about whether to pursue a global thermostat has to be political before it can be scientific. It is not a case of researching into the risks and benefits of the technology to begin with, and then subjecting the results to a risk calculus. Nor is it sufficient to investigate different governance options. The priority must be to reveal the beliefs, ideologies, values and interests of those who are promoting the technology and to argue about why it should even be contemplated. One needs to introduce public and political deliberations further 'upstream' before prospective research is even mandated or funded, and to deliberate upon this particular technological intervention in the context of all other possible responses to climate change. One needs to give voice to a multitude of arguments about why it may not be desirable to embark even on the path of research and development. And voices from around the world must be heard – not just from the elites of Anglophone nations – or of richer nations in general. Otherwise we risk entering a new era of tyranny, and the mighty fascism of naturalism will suppress the creative tension and moral responsibility of agonistic human beings.

Although I am opposed to researching the aerosol injection technologies that might enable a thermostat for the planet, it should be clear that I am not antagonistic to the idea of science and technology being harnessed and used, in more modest and less ambitious ways, to reduce the risks and dangers of a changing climate. I have endorsed an approach called climate pragmatism, which puts science and technology into a framework of action different from the hubris and megalomania of stratospheric aerosol injection. The global thermostat is an inadequate solution to a wrongly framed problem. It is more important to pursue clumsy solutions to well-chosen problems than to seek out universal solutions to wicked problems.[58]

The approach outlined draws attention to the virtues of *pluralism* and *pragmatism*. Pluralism offers checks and balances to scientific and political projects alike. It recognises the inevitability of competing values and goals and concedes that a world of more than 7 billion people cannot move together. Such a world will not agree on a single thermostat setting. The corollary of pluralism is philosophical and political pragmatism. Philosophical pragmatism forgoes the goal of establishing an ultimate truth in favour of working with merely satisfactory truths, while political pragmatism suggests a cautious and flexible approach to defining the problems we decide to tackle. Pragmatism is thus content to recognise and name problems like climate change as being super-wicked in character: not definable and not solvable. Instead of using science and technology to 'fix' wicked problems, pragmatism is content to pursue multiple and clumsy solutions to regularly reframed problems in order to achieve merely incremental gains.

This approach has been articulated in recent years most clearly by the Hartwell Group and by the Breakthrough Institute. It transforms the singular problem of climate change – which the thermostat is imagined to fix – into a series of differently bounded problems. Climate pragmatism promotes investment in social and infrastructural adaptations designed to reduce weather risks where such reduction is most needed. It lends weight to regulatory and technical controls on emissions of a range of non-carbon dioxide pollutants that damage the health of humans and of ecosystems as well as altering the climate. And it advocates the use of science and engineering to create affordable large-scale low-carbon or zero-carbon energy technologies designed to meet the huge but still growing world demand for energy. These more modest and more limited goals enlist science and technology in the service of human(e) values rather than being offered up for hubristic projects. They do not seek to fix climate change. They do not seek to stabilise global temperature at two degrees. But climate pragmatism *does* seek to advance human welfare and to meet the needs of development that are inscribed in basic notions of dignity and fairness.

Aldous Huxley's 1932 novel *Brave New World* can be understood as portraying either a too perfectly controlled utopia or a brutal and inhuman dystopia. It offers a way of imagining the consequences of seeking to perfect humanity through ever increasing technological intrusions. But in this brave new world human enhancement comes at a terrible cost. I compare Huxley's tale with the vision behind the technology of stratospheric aerosol injection and other global sunlight reflection methods. These thermostat visionaries are seeking to enhance the world's climate in the name of a noble cause. They offer technologies that will not just compensate for a disturbed global heat balance; they will end up doing far more. They will inaugurate an era of never ending experimentation with the global sky. In their search for ever more control and adjustment, they will convert the world's climate – all our climates – into something relentlessly unnatural. Stratospheric aerosol injection will usher in a brave new world of designer climates.

There is a form of experimentation that is both inevitable and desirable for humans, given their insatiable curiosity and appetite for creativity. But, since we are inescapably enlisted on the project of remaking the world, I would rather do so slowly, humbly and, yes, accidentally. Let us attend to the difficult pursuit of liberty, justice and human security on the ground and not delude ourselves that utopias can be engineered in the sky.

[1] The definition of the phrase 'climate change' is problematic – and its meaning has changed over time. For the purposes of this book I use the idea of 'climate change' to embrace the physical change (and, within this concept, both natural and human-caused change) *and* the cultural dimensions of the phenomenon, together with their interrelationships.

[2] Garrard 2013: 183.

[3] London: Chatto and Windus, 1932.

[4] McEwan 2010: 213.

[5] Crutzen 2006.

[6] Kintisch 2007: 1054.

[7] Ibid.

[8] Ibid., 1055.

[9] Olson 2012.

[10] Fleming 2010: 77–8.

[11] President's Science Advisory Committee 1965: 127.

[12] As summarised in Brahic 2009.

[13] ABC News 2009.

[14] Quoted in Chandler 2007: 42.

[15] Welch, Gaines, Marjoram and Ionesca 2012.

[16] Hamilton 2013a.

[17] Calhoun 2008: 376.

[18] Ibid., 375.

[19] Saunders 2009.

[20] Arctic Methane Emergency Group n.d.

21 Bickel 2013: 2.
22 Ken Caldeira speaking in 2010, quoted in Sikka 2012: 168.
23 Calhoun 2008: 391.
24 Heyward and Rayner 2013.
25 Pleij 2001: 180–1.
26 de Buffon 1778: 244.
27 Meyer 2002: 593.
28 Anonymous 1845: 174–5.
29 Johnson and Murphy 1837: 367.
30 Scott 1998.
31 Ricke, Morgan and Allen 2010: 537.
32 Von Neumann 1955: 43.
33 Ricke, Moreno-Cruz and Caldeira 2013: 6.
34 Scott 2012: 134.
35 Crowley 2011.
36 Engineering and Physical Sciences Research Council 2011 and 2012.
37 Michaelson 1998.
38 See Rayner et al. 2013.
39 Parson and Keith 2013.
40 Hamilton 2013b: 139.
41 Rapp 1989.
42 Fleming 2010.
43 Hands Off Mother Earth 2011.
44 Big Dead Place n.d.
45 Hands Off Mother Earth 2011.
46 Revelle and Suess 1957: 19.
47 See www.margaretthatcher.org/document/107346.
48 O'Neill 2012.
49 Marglin, quoted in Scott 1998: 344.
50 Myles Allen, quoted in Gillis 2011.
51 Macnaghten and Szerszynski 2013: 470.

[52] Preston 2012a: 198.

[53] This illustration was first used in connection with climate change policy by Steve Rayner and is found in Prins and Rayner 2007.

[54] Parts of this section are based on Chapter 10 of M. Hulme (2009) *Why We Disagree About Climate Change: Understanding Controversy, Inaction and Opportunity*. Cambridge: Cambridge University Press.

[55] See Prins et al. 2010; Atkinson et al. 2011.

[56] Snow 1998: 46 (emphasis added).

[57] Lovelock 2010.

[58] Rayner 2006.

ABC News. (2009) Obama considers climate engineering to cool globe, 8 April, http://abcnews.go.com/Technology/story?id= 7313752 (accessed 19 December 2013).

Anonymous. (1845) Young America: Anti-rentism. *The Harbinger*, 23 August, pp. 174–5.

The Arctic Methane Emergency Group (AMEG). (n.d.), http://www .ameg.me/index.php/emergency (accessed 14 August 2013).

Atkinson, R., Chhetri, N., Freed, J., Galiana, I., Green, C., Hayward, S., Jenkins, J., Malone, E., Nordhaus, T., Pielke Jr, R., Prins, G., Rayner, S., Sarewitz, D. and Shellenberger, M. (2011) *Climate Pragmatism: Innovation, Resilience and No Regrets: The Hartwell Analysis in an American Context*. Washington, DC: Breakthrough Institute.

Bickel, J. E. (2013) Climate engineering and climate tipping point scenarios. *Environmental System Dynamics* 33: 151–67.

Big Dead Place. (n.d.) Embracing the experiment: Interview with Abigail Alling of Biosphere 2, http://www.bigdeadplace.com/ stories-and-interviews/embracing-the-experiment-interview-with -abigail-alling-of-biosphere-2/ (accessed 4 September 2013).

Brahic, C. (2009) Earth's Plan B. *New Scientist*, 28 February, pp. 8–10.

Burns, W. C. G. and Strauss, A. L. (eds.) (2013) *Climate Change Geoengineering: Philosophical Perspectives, Legal Issues, and Governance Frameworks*. Cambridge: Cambridge University Press.

Bibliography

Calhoun, C. (2008) A world of emergencies: Fear, intervention and the limits of cosmopolitan order. *Canadian Review of Sociology* 41(4): 373–95.

Chandler, D. (2007) A sunshade for the planet. *New Scientist*, 21 July, pp. 42–5.

Crowley, T. (2011) I hope we never need geoengineering, but we must research it. *The Guardian*, 8 September, http://www.theguardian.com/environment/2011/sep/08/geoengineering-research-royal-society (accessed 2 September 2013).

Crutzen, P. (2006) Albedo enhancement by stratospheric sulfur injections: A contribution to resolve a policy dilemma. *Climatic Change* 77(3/4): 211–20.

de Buffon, G.-L. L. (1778) *Histoire naturelle, générale et particulière, vol. 20: Époques de la nature.* Paris: Imprimerie royale.

Engineering and Physical Sciences Research Council. (2011) Website, 29 September, http://www.epsrc.ac.uk/newsevents/news/2011/Pages/spiceupdate.aspx (accessed 2 September 2013).

Engineering and Physical Sciences Research Council. (2012) Website, 17 October, http://www.epsrc.ac.uk/newsevents/news/2012/Pages/spiceupdateoct.aspx (accessed 2 September 2013).

Fleming, J. R. (2010) *Fixing the Sky: The Checkered History of Weather and Climate Control.* New York: Columbia University Press.

Garrard, G. (2013) The unbearable lightness of green: Air travel, climate change and literature. *Green Letters: Studies in Ecocriticism* 17(2): 175–88.

Goodell, J. (2010) *How to Cool the Planet: Geoengineering and the Audacious Quest to Fix Earth's Climate.* Boston, MA: Houghton Mifflin Harcourt.

Hamilton, C. (2013a) *Earthmasters: The Dawn of the Age of Climate Engineering.* New Haven, CT/London: Yale University Press.

Hamilton, C. (2013b) No, we should not just 'at least do the research'. *Nature* 496: 139.

Bibliography

Hands Off Mother Earth (HOME). (2011) Open letter to IPCC on geoengineering, 13 June, http://www.handsoffmotherearth.org/2011/06/lettertoipcc/ (accessed 2 September 2013).

Heyward, C. and Rayner, S. (2013) Apocalypse nicked: Climate Geoengineering Governance Working Paper No. 6. Oxford: James Martin School, http://geoengineering-governance-research.org/perch/resources/workingpaper6heywardraynerapocalypsenicked.pdf (accessed 1 October 2013).

Gillis, J. (2011) Heavy rains linked to humans. *New York Times*, 16 February, www.nytimes.com/2011/02/17/science/earth/17extreme.html?_r=1 (accessed 5 September 2013).

Hulme, M. (2009) *Why We Disagree about Climate Change: Understanding Controversy, Inaction and Opportunity*. Cambridge: Cambridge University Press.

Johnson, S. and Murphy, A. (1837) *The works of Samuel Johnson, LL.D., with an essay on his life and genius*, vol. 1. New York: George Dearborn.

Keith, D. (2013) *A Case for Climate Engineering*. Cambridge MA: MIT Press.

Kintish, E. (2007) Scientists say continued warming warrants closer look at drastic fixes. *Science* 318: 1054–5.

Kintisch, E. (2010) *Hack the Planet: Science's Best Hope – or Worst Nightmare – for Averting Climate Catastrophe*. Hoboken, NJ: Wiley & Sons.

Long, J., Rademaker, S., Anderson, J. G., Caldeira, K., Chaisson, J., Goldston, D., Hamburg, S., Keith, D., Lehman, R., Loy, F., Morgan, G., Sarewitz, D., Schelling, T., Shepherd, J., Victor, D. G., Whelan, D. and Winickoff, D. E. (2011) *Task Force on Climate Remediation Research*. Washington, DC: Bipartisan Policy Center.

Lovelock, J. (2010) Fudging data is a sin against science. *The Guardian*, 29 March, http://www.theguardian.com/environment/2010/mar/29/james-lovelock (accessed 20 September 2013).

Bibliography

McEwan, I. (2010) *Solar*. London: Jonathan Cape.

Macnaghten, P. and Szerszynski, B. (2013) Living the global social experiment: An analysis of public discourse on geoengineering and its implications for governance. *Global Environmental Change* 23(2): 465–74.

Marchetti, C. (1977) On geoengineering and the CO_2 problem. *Climatic Change* 1(1): 59–68.

Meyer, W. B. (2002) The perfectionists and the weather: The Oneida community's quest for meteorological utopia, 1848–1879. *Environmental History* 7(4): 589–610.

Michaelson, J. (1998) Geoengineering: A climate change Manhattan Project. *Stanford Environmental Law Journal* 17(73): 74–140.

O'Neill, J. (2012) The ethics of engineering. *Environmental Values* 21(1): 1–4.

Olson, R. L. (2011) *Geoengineering for Decision-Makers: Science and Technology Innovation Program*. Washington, DC: Woodrow Wilson International Center for Scholars.

Olson, W. W. (2012) Soft geoengineering: A gentler approach to addressing climate change. *Environment* 54(5): 29–39.

Parson, E. A. and Keith, D. W. (2013) End the deadlock on governance of geoengineering research. *Science* 339: 1278–9.

Pleij, H. (2001) *Dreaming of Cockaigne: Medieval Fantasies of the Perfect Life*, trans. D. Webb. New York: Columbia University Press.

President's Science Advisory Committee. (1965) *Restoring the Quality of Our Environment*. Report of the Environmental Pollution Panel, Washington, DC.

Preston, C. J. (2012a) Beyond the end of nature: SRM and two tales of artificity for the Anthropocene. *Ethics, Policy and Environment* 15(2): 188–201.

Preston, C. J. (ed.). (2012b) *Engineering the Climate: The Ethics of Solar Radiation Management*. Lanham, MD: Lexington Books.

Bibliography

Prins, G. and Rayner, S. (2007) *The Wrong Trousers: Radically Re-Thinking Climate Policy.* London: London School of Economics, Oxford: James Martin Institute for Science and Civilisation.

Prins, G., Galiana, I., Green, C., Grundmann, R., Hulme. M., Korhola, A., Laird, F., Nordhaus, T., Pielke Jr, R., Rayner, S., Sarewitz, D., Shellenberger, M., Stehr, N. and Tezuka, H. (2010) *The Hartwell Paper: A New Direction for Climate Policy after the Crash of 2009.* London: London School of Economics.

Rapp, F. (1989) Introduction: General perspectives on the complexity of philosophy of technology. In P. T. Durbin (ed.), *Philosophy of Technology: Practical, Historical, and Other Dimensions.* Dordrecht and Boston, MA: Kluwer Academic, pp. ix–xxiv.

Rayner, S. (2006) Wicked problems: Clumsy solutions: Diagnoses and prescriptions for environmental ills, http://www.insis.ox.ac.uk/fileadmin/InSIS/Publications/Rayner_-_jackPeaPePecture.pdf.

Rayner, S., Heyward, C., Kruger, T., Pidgeon, N., Redgwell, C. and Savulescu, J. (2013) The Oxford Principles for Geoengineering Governance. *Climatic Change* 121: 499–512.

Revelle, R. and Suess, H. E. (1957) Carbon dioxide exchange between atmosphere and ocean and the question of an increase of atmospheric CO_2 during the past decades. *Tellus* 9: 18–27.

Ricke, K. L., Moreno-Cruz, J. B. and Caldeira, K. (2013) Strategic incentives for climate geoengineering coalitions to exclude broad participation. *Environmental Research Letters* 8(1): 1–8 (014021). DOI: 10.1088/1748-9326/8/1/014021

Ricke, K. L, Morgan, M. G. and Allen, M. R. (2010) Regional climate response to solar-radiation management. *Nature Geoscience* 3(8): 537–41.

Royal Society. (2009) *Geoengineering the Climate: Science, Governance and Uncertainty.* London: Royal Society.

Saunders, A. (2009) The dangers of geo-engineering. *The Guardian*, 1 March, http://www.theguardian.com/commentisfree/

2009/feb/27/climate-change-geo-engineering (accessed 14 August 2013).

Scott, D. (2012) Introduction to the special section, 'The ethics of geoengineering: Investigating the moral challenges of solar radiation management'. *Ethics, Policy and Environment* 15(2): 133–5.

Scott, G. C. (1998) *Seeing like a State: How Certain Schemes to Improve the Human Condition Have Failed.* New Haven, CT: Yale University Press.

Sikka, T. (2012) A critical discourse analysis of geoengineering advocacy. *Critical Discourse Studies* 9(2): 163–75.

SMRGI. (2011) *Solar Radiation Management: The Governance of Research.* London: EDF/Royal Society/TWAS.

Snow, C. P. (1998) *The Two Cultures (with Introduction by S. Collini).* Cambridge: Cambridge University Press.

UK Parliament. (2010) *The Regulation of Geoengineering: House of Commons Select Committee on Science and Technology (HC 2210).* London: HMSO.

US Government Accountability Office. (2010) *Engineering the Climate: Research Needs and Strategies of International Coordination.* Report for 111th Congress, Second Session. Washington, DC.

US Government Accountability Office. (2011) *Climate Engineering: Technical Status, Future Directions and Potential Responses.* GAO-11-71. Washington DC.

von Neumann, J. (1955) Can we survive technology? *Fortune* 91(6): 32–47.

Welch, A., Gaines, S., Marjoram, T. and Fonesca, L. (2012) Climate engineering: The way forward. *Environmental Development* 2: 57–72.

Index

Index

Index

Index